Railway Company Union Pacific

A complete and comprehensive description of the

agricultural, stock raising and mineral resources of Utah

Railway Company Union Pacific

A complete and comprehensive description of the agricultural, stock raising and mineral resources of Utah

ISBN/EAN: 9783744726801

Printed in Europe, USA, Canada, Australia, Japan

Cover: Foto ©berggeist007 / pixelio.de

More available books at **www.hansebooks.com**

A COMPLETE AND COMPREHENSIVE DESCRIPTION

OF THE

AGRICULTURAL, STOCK RAISING

—AND—

MINERAL RESOURCES

—OF—

UTAH

WITH THE COMPLIMENTS OF THE

PASSENGER DEPARTMENT

Sixth Edition, Revised and Enlarged.

ST. LOUIS.
WOODWARD & TIERNAN PRINTING CO., 309-319 NORTH THIRD STREET.
1893.

INTRODUCTORY.

Before proceeding with a presentation of the resources, attractions and peculiar features of Utah—the most populous and thrifty of the remaining Territories—we deem it proper to say a few words regarding the great highway which first connected her with the commercial world, and in bonds of steel united her to the great Sisterhood—the UNION PACIFIC RAILWAY.

When the building of this national Giant's Causeway was begun, all the domain west of a narrow strip bordering the Missouri River and east of Utah, with the exception of a cultivated spot here and there, was a barren waste. The somewhat uninviting spectacle of a country contributing little or no patronage, until it could be created by the road itself, for the distance of a thousand miles, was thus presented, the chief reliance of the Company being the through business that must be poured upon it from the congested communities of the Far West when the road was completed, and the settlement and development of the country through which it passed. The road would make the latter a feasible project, with the accomplishment of which mutual benefits to the settlers and the Company would result; these, expanding and increasing from time to time, as civilization was extended further and further into the interior, and the villages took on more and more the characteristics and conditions of towns, and these of cities. It was a grand project. It meant the practical annihilation of the American frontier and the occupation and subjugation of millions of acres of the best soil in the world, and which previously was used for no other purpose than as a roaming ground for wild beasts and wilder man. Look at it now, with scarcely half a generation gone since ground was broken and the great evangel of progress began its majestic and all-conquering march into the wilderness! Where once the waters of the mountains and plains aimlessly meandered to the rivers which feed the ocean, are now diverting channels conveying the life-giving element over fertile and productive fields, into gardens and orchards, and through the streets and by-ways of prosperous and growing communities; the "lowing herds wind slowly o'er the lea" where formerly the bison led an aimless life; the war-whoop of the savage is heard no more, and in its place the clangor of machinery and the "sound of the church-going bell" are wafted upon the breezes; and wherever we may turn, "civilization, on its luminous wings, soars phœnix-like to Jove." The land from center to circumference has undergone a metamorphosis so complete and wonderful that it seems to have been wrought by magic—and so, indeed, it was, the magic of the locomotive trained by hands of skill and directed by the finger of American enterprise.

The country thus reclaimed and made habitable to man amounts to an empire in extent and wealth. Great as it already is, its greatness is but just begun. But a limited fraction is yet taken up and homes for millions yet remain. What a grand field for the poorly-paid toiler in the crowded cities of the East, to acquire independence at once and wealth in the immediate future. It requires but small capital with enterprise, intelligence and industry, to effect the complete disenthralment of the homeless wage-worker or unfortunate plodder; the fruitful soil and the mountains teeming with undeveloped wealth are accessible to him, while incurring no risks from savages or the other dangers that constantly confronted settlers who had the temerity to venture beyond the border in the early days.

Those who read the interesting chapters following will be advised of what has been wrought out of just such materials as the Union Pacific Railway traverses, almost from end to end, to reach the progressive and fruitful Territory treated of. It all, in contemplation of the past, reads like a production of Munchausen instead of, as it is, the recital of accomplished realities well worth going to see at once. The Union Pacific Railway Company wishes every patron a liberal share of all these good things in Utah. In attracting attention to the well-filled pages of this little work, it also desires to say that it is the result of much labor on the part of the very best authorities. Special acknowledgment is made to S. A. Kenner, for many years a journalist and member of the bar of Salt Lake City, for the preparation of the pamphlet.

CONTENTS.

UTAH TERRITORY.

GRAND PHYSICAL FEATURES.

AREA.—Utah Territory is in the latitude of Missouri, about two-thirds of the way from St. Louis to San Francisco. Its land area is 84,970 square miles (52,601,600 acres); its water area is 2,780 square miles (1,776,200 acres). The grand physical features of Utah are as follows:

DIVISIONS.—The Wasatch Mountains and the High Plateaus, a central zone standing 9,000 to 11,000 feet above the sea, and extending from the northern nearly to the southern boundary; the Uintah Mountains, a lofty table land, carrying many peaks 12,000 to nearly 14,000 feet high, stretching eastward from the middle of the Wasatch; the Tavaputs Table Land, another elevated district, south from the Uintah region, extending from the southern extremity of the Wasatch Mountains east-southeast beyond the borders of Utah, and cut in two by Green River; the Uintah-White Basin, a low synclinal valley, drained by the Uintah and its tributaries on the west, and the lower course of White River on the east, lying between the Uintah Mountains and the Tavaputs Table Land; the Cañon Lands, south of the Tavaputs Table Land and east and south of the High Plateaus, in the midst of which the Green and the Grand unite to form the River Colorado, and which is traversed in deep cañons by the Price and the San-Rafael, by the Fremont, the Escalante, the Paria, the Kanab, and the Virgin Rivers and their tributaries; the Great Basin, subdivided into the Sevier Lake Basin and the Great Salt Lake Basin, a region of low valleys lying west of the lofty zone. interrupted by short and abrupt ranges of mountains, part of a mountain system extending through Nevada and northwestward into Idaho and Oregon.

DRAINAGE SYSTEMS.—The eastern part of the Territory is drained by the Rio Colorado and its tributaries; the western part by streams that head in the Wasatch and the High Plateaus of the central part, and find their way into the salinas and desert sands of the Great Basin. Thus we have the Rio Colorado drainage area, and the Desert drainage area; the former about two-fifths, the latter about three-fifths, of the total area. The Rio Colorado drainage area is subdivided into the Uintah-White

(7)

Basin, with 280,320 acres of irrigable land, and the Cañon Lands, with 213,440 acres. The Desert drainage area is subdivided into the Sevier Lake District, with 101,700 acres of irrigable land, and the Great Salt Lake District, with 837,660 acres. *

GREAT SALT LAKE DRAINAGE SYSTEM.—Three rivers enter Great Salt Lake, namely, the Bear, the Weber, and the Jordan, "and upon their water," says Mr. C. K. Gilbert, of the Geological Survey, "will ultimately depend the major part of the agriculture of Utah." They rise close together in the western end of the Uintah Mountains, and cut through the Wasatch. Bear River flows northward, now in Utah, now in Wyoming, and into Idaho as far as Soda Springs. Here it bends round like a fish hook and returns on a more westerly line. Re-entering Utah in Cache Valley, it passes thence by a short cañon to its delta-plain on the northeastern border of Great Salt Lake. Its principal tributaries are received in Idaho and Cache Valley.

Cache Valley, in Utah and Idaho, contains upwards of 400 square miles of irrigable land. The left bank (of the Bear) is served by Logan River and tributaries; the right bank by a canal (not yet constructed) entirely in Idaho. The expense of the latter will be great, but well warranted. The valley is higher and somewhat colder than the Salt Lake Valley, but the soil is good, and the climate admits of the growth of wheat, oats, corn, rye, apples, pears, cherries, apricots, plums, peaches, etc. The valley is about ten miles in width by fifty miles in length, dish-like in shape, walled in by mountains, but pretty well farmed all around at the foot of the mountains. It sustains nearly a score of flourishing towns.

The mean annual flow of Bear River, where it enters Salt Lake Valley, is about 5,000 cubic feet per second. Its delta-plain contains about 220 square miles of unsurpassable soil, upon which the Bear River Canal Com any has diverted 2,000 second-cubic feet of water through upwards of 100 miles of canals at a cost of nearly $3,000,000. The soil is rich and ideally adapted to irrigation, having a gentle fall, being smooth as a floor, and well and deeply drained by the Bear and Malad Rivers.

As if to forever bar a water famine in Salt Lake Valley, nature has provided a natural reservoir in Bear Lake, situated near Bear River and connected with the river by a narrow outlet, high up in the mountains. The lake has an area of 150 square miles, and can be raised ten feet by a dam thrown across the outlet at slight expense. Thus enough water can be stored during three-fourths of the year to flow 5,000 feet per second during the other fourth of the year. Bear River itself can be turned into the lake by a short canal, and upon its upper tributaries, on the northern slope of the Uintah Range, there are many glacial lakes which can be made use of for impounding water.

---

*NOTE.—This matter is condensed from Major Powell's Lands of the Arid Region, Government Printing Office, Washington, 1879. Further investigation indicates that the irrigable land was at that time considerably under-estimated.

The Weber River runs in a general northwesterly course from the Uintah Mountains to Great Salt Lake, entering the latter at the middle of its eastern shore. The Ogden is its only important tributary. Its delta-plain comprises about 220 square miles of farming land. If the river prove incompetent to water its delta-plain, the Bear at the north and the Jordan at the south have each a great volume of surplus water, and either supply can be led without difficulty to the lower levels of the delta of the Weber. Besides the delta of the Weber, there are 40 to 50 square miles of irrigable land on the Weber and the Ogden Rivers within the mountains.

The Jordan River is the outlet of Utah Lake, and runs northward, entering Great Salt Lake at its southeastern angle. On the right it receives a number of large tributaries from the Wasatch. The largest tributary of Utah Lake is Provo River, which rises near the source of the Weber and the Bear in the Uintah Mountains. Minor tributaries of Utah Lake are American Fork, Spanish Fork, Hobble Creek, Payson Creek, Salt Creek, e c. On all the tributaries of Utah Lake there are about 320 square miles of irrigable land; and in Jordan Valley, below Utah Lake, inclusive of Bountiful and Centerville, there are about 250 square miles. In addition, the water can be carried around the point of the Oquirrh Range on the southern shore of Great Salt Lake, and be used to water fifty square miles in Toeele Valley.

Utah Lake is a natural reservoir, 125 square miles in surface area. With suitable headworks its volume can be controlled, and the entire discharge be concentrated in the season of irrigation. The mean volume of the outlet is about 1,000 second-cubic feet, but one-fourth of this must be assigned to watering lands on the tributaries of the lake and to evaporation, leaving a perennial flow of 750 second-cubic feet, which, if concentrated into four months, would irrigate for that period 350 square miles.

There is thus water enough forever assured to irrigate every acre of the eastern border of Great Salt Lake Basin, from Nephi on the south to Bear River Cañon on the north, a distance, as traveled, of about 180 miles. This fringe of the desert, between the Wasatch and Great Salt Lake, and between the Wasatch and Utah Lake, is, in location, resources, climate fertility, potentially the glory of the earth. It is easily the garden spot of Utah. It supports more than thirty settlements or towns, and more than half the population of Utah. Every acre of the land is intrinsically worth $100, although it ranges in price all the way from $5 to $225 per acre. The average, away from the suburbs of larger towns, is perhaps $50 an acre. Altogether, about 10,000 second-cubic feet of water perennially flows into Great Salt Lake.

Westward of Great Salt Lake there are sixty small tracts of land blest with water. On the east of the lake the rivers carry the melting snows of the elevated zone to the valleys, and fertility is the result. West and north of the lake the mountains are too insignificant to store up snow banks until the time of need. These streams are spent before the summer comes, and only a few springs are perennial. The result is a desert, with little oases a day's journey apart.

SEVIER LAKE DRAINAGE BASIN.—According to the accomplished geologists of the United States Geological Survey, which this sketch follows, the Wasatch ends with Mount Nebo, which overhangs Nephi. The elevated lands southward these gentlemen term the High Plateaus, divided by great longitudinal faults into three ranges, each made up of different members, as the San Pete, the Pahvant, the Tushar, and the Markagunt, facing the Great Basin; the Sevier and the Paunsagunt between Sevier and Grass valleys; and the Wasatch, the Fish Lake, the Awapa, and the Aquarius, east of Grass Valley. The Pahvant and the Tushar, says Captain Dutton, present a curious admixture of plateau and ˙sierra, but the others are true tables, made and kept so by the lavas which cap them and successfully resist erosion.

The Wasatch Plateau is east of San Pete Valley, above which it rises a full mile. Sanpitch River, the largest tributary of the Sevier, furnishes water, and the oats and wheat grow higher than the fences. There is coal in the valley, fine building and flagging stone, a score of towns and settlements, and 50,000 to 100,000 acres of irrigable land. The Sanpitch empties into the Sevier at Gunnison, the latter coming down from the south, the former rising about Mount Nebo and flowing southward.

From Gunnison to Monroe, Sevier Valley is about five miles wide by sixty miles long, and sustains a dozen settlements. The river cañons above Monroe, and just above this cañon tower the rugged peaks and domes of the Tushar (Beaver Range), upon whose shaggy slopes, descending to the Sevier, is the mining district of Marysvale, just now rousing itself, or being roused, from a Rip Van Winkle sleep of 20 years.

Twenty miles above Marysvale is Circle Valley, where the East Fork joins the South Fork through a mighty chasm, cutting the Sevier Plateau in two. The mural walls of the opposing plateaus rise sheer above Circle Valley 4,000 to 5,000 feet. From this junction the two forks continue on through cañons and valleys, ascending higher and higher the best part of a hundred miles to the springs of the basalt fields which divide the drainage of Sevier Lake from that of the Rio Colorado. There are valleys up there, says Captain Dutton, 7,000 to 9,000 feet high, with the palisades of the Plateaus rising half a mile higher, and on the great mesas forests of straight slender pines and spruces a hundred feet to a limb and standing so thick as to be almost impenetrable.

Just below Juab the Sevier River breaks through the Pahvant as though the latter were a fog-bank, runs far out on the desert and sinks in what is called Sevier Lake. Without storage, for which Captain Dutton says the High Plateaus offer extraordinary facilities, the Sevier and Sanpitch rivers water less than 100,000 acres. With storage, if there is sufficient water to be stored, a thousand square miles of land might be reclaimed from the desert on the course of the Sevier River.

Probably a hundred square miles are served by the small streams of southwestern Utah, as at Levan, Scipio, Holden, Fillmore, Oak City, Kanosh, Beaver, Minersville, Paragoonah, Parowan, Cedar City, Pinto, Hebron, etc. In this region the water is inadequate to supply the arable land, but it can be largely increased by storage, without doubt.

COLORADO RIVER DRAINAGE.—Of the Rio Colorado drainage system, the main channel is the river Colorado and its proper continuation, the Green River. The principal tributaries of these streams from the east are the White, the Grand, and the San Juan, the White entering the Green, the Grand uniting with the Green to form the Colorado, and the San Juan entering the latter about 125 miles below the confluence of the Grand and the Green. The tributaries from the west are the Virgin, the Kanab, the Paria, the Escalante, the Fremont, the San Rafael, the Price, the Minnie Maud, the Uintah, and the Ashley Fork.

The climate is extremely arid, the elevation between 2,500 and 11,500 feet, giving great range in temperature. The limit of successful (hay) farming is about 7,000 feet. Aside from the Uintah-White Basin, which contains more than half of the irrigable land of the entire district, and which is an Indian Reserve, the lands are generally on benches or terraces or in restricted valleys between the higher courses of the streams and their cañons, and from 4,500 to 6,000 feet in altitude. The Price, the Uintah, the Green and the Grand have plenty of water, but, excepting the Uintah, the land upon which their waters can be diverted is very limited. On the Virgin, which is far south and low in altitude, there are thirty to fifty square miles. In the entire district there may be a thousand square miles of irrigable-arable land.

AGRICULTURAL, TIMBER, AND GRAZING LANDS.—From a cursory examination and estimate of the water supply, made under Major Powell's auspices in 1877, the land in Utah which may be irrigated was tentatively put at 1,433,000 acres. Later and more thorough investigation indicates at least 3,000,000 acres.

Upon the high mountain slopes and mesas are the forests. All the timber trees proper are coniferous and belong to the Pine, Fir, and Juniper families. There will doubtless always be enough timber and lumber for domestic use, as the new growth should replace the consumption. The farming lands, on the lower courses of the rivers and near the mountains, are limited in extent, and coal is so plentiful as to be universally used for fuel. No timber or lumber should ever be exported from Utah, nor are they likely to be. Major Powell estimates the timber region at 18,500 square miles; standing timber at 10,000; milling timber at 2,500 square miles; sufficient, he says, for the industrial wants of the country if it can be preserved from forest fires.

The elevated regions not only store the moisture to fertilize the adjacent lowlands, but they contain the mines of silver and gold, of lead and iron, and of other metals and minerals, and the coal.

The grazing lands lie in the main between the high timber lands and the low farming lands. The grass is scanty, but in great variety and nutritious.

Wherever grass grows, Major Powell says, water may be found, or saved from the rains in sufficient quantity for all the herds that can live on the pasturage.

GEOLOGY; GEOLOGICAL HISTORY.

FORMATION.—In Clarence King's "Exploration of the Fortieth Parallel" occurs the following:

"The greater part of the rock of the interior mountain area is a series of conformable stratified beds reaching from the early Azoic to the late Jurassic times. In the latter these beds were raised—the Sierras, the Wasatch, and paralleled ranges of the Great Basin were the consequence. In this upheaval important masses of granite broke through, accompanied by quartz, porphyries, feldsite rocks, and notably sienitic granite with granulite and gretsen occasionally. Then the Pacific Ocean on the west, and the ocean that filled the Mississippi basin on the east, laid down a system of Cretaceous and Tertiary strata. These outlying shore beds, subsequently to the Miocene, were themselves raised and folded, forming the Pacific Coast Range and the chains east of the Wasatch; volcanic rocks accompanying this upheavel as granite did the former one. Still later a final series of disturbances occurred, but these had but small connection with the region under consideration.

"There is a general parallelism of the mountain chains, and all the structural features of local geology, the ranges, strike of great areas of upturned strata, larger outbursts of gigantic rocks, etc., are nearly parallel with the meridian. So the precious metals arrange themselves in parallel longitudinal zones. There is a zone of quicksilver, tin, and chromic iron on the Coast Range, one of copper along the foot-hills of the Sierras; one of gold further up the Sierras—the gold veins and resultant placers extending far into Alaska; one of silver with comparatively little base metal along the east base of the Sierras, stretching into Mexico; silver mines with complicated associations through Middle Mexico, Arizona, Middle Nevada and Central Idaho; argentiferous galena through New Mexico, Utah and Western Montana; and, still further east, a continuous chain of gold deposits in New Mexico, Colorado, Wyoming and Montana. The Jurassic disturbance in all probability is the dating point of a large class of lodes: a, those wholly enclosed in the granites, and b, those in metamorphic beds of the series extending from the Azoic to the Jurassic. To this period may be referred the gold veins of California, those of the Humboldt mines, and those of White Pine, all of class b; and the Reese River veins, partly a and partly b. The Colorado lodes are somewhat unique, and in general belong to the ancient type. To the Territory period may be definitely assigned the mineral veins traversing the early volcanic rock; as the Comstock Lode and veins of the Owyhee District, Idaho. By far the greater number of metalliferous lodes occur in the stratified metamorphic rocks or the ancient eruptive rocks of the Jurassic upheaval; yet very important, and, perhaps, more wonderfully productive, have been those silver lodes which lie wholly in the recent volcanic formations."

THE GREAT BASIN.—Major Powell, head of the U. S. Geological Survey, holds that what is now the Great Basin was the first part of the West to emerge from the sea. During the whole of Mesozoic time the Great

Basin was drained into a sea which covered the Wasatch, the Uintahs, and the High Plateaus—a sea in which the lofty mountains enclosing the great parks of Colorado were a chain of islands. At the close of the Mesozoic or the opening of the Cenozoic, the uplifting of the continent between St. Louis and San Francisco began; the surface of the land was broken, contorted and distorted along innumerable lines; the Great Plains and the Great Basin were raised to about their present altitude above the sea, while the Sierra Nevada and the Sierra Madre, the Wasatch, the Uintah and other great chains, and that vast region of plateaus and cañons drained by the Colorado River of the West, were slowly carried on upward a mile, two miles, ten miles higher, until the drainage was turned into the Great Basin, which had formerly been out from it. That which before was lofty was abased, and that which before was abased was reared aloft. This vast movement, or rather series of movements, is supposed to be still going on. At times and in places it has been violent and accompanied by great volcanic activity, but in general it has probably proceeded almost imperceptibly, and may even have been quiescent for long periods. In Millard and Kane counties the evidences of volcanic or seismic action, or both, are most distinct.

The ranges of the Great Basin, Major Powell holds to be of very recent origin, probably as recent as late Tertiary. These ranges he describes as "short, more or less distinct, north and south ridges, separated by desert valleys, which reveal broad stretches of sub-aerial gravels concealing the underlying formations." These ranges are usually simple monoclinal ridges, produced by the up-tilting of comparatively rigid bodies of strata along one side of a vertical or nearly vertical fault, suggesting the application of vertical pressure from below. Volcanic phenomena abound and are intimately associated with the ridges of upheaval. The edges of these upheaved blocks have been worn into rugged, bristling and sierra-like forms, and modified by flows of eruptive matter from below. The ridges are composed of Paleozoic rocks, with Azoic schists beneath. The eruptive rocks appear upon the flanks of the ridges, sometimes partly masking them, again cutting them transversely or obliquely, and in many ways complicating the topographic structure. From the fact that these ridges have not been eroded far back from their lines of faults, Major Powell concludes that they were never greatly upheaved beyond their present altitudes. Still, each ridge is but a small residuary fragment of the great inclined block, and the inter-range spaces are filled with clays, sands, and gravels, the waste of these blocks, in such manner as to bury the underlying rocks over broad areas. The amount of this transferred material is very great.

THE WASATCH AND THE HIGH PLATEAUS.—The Wasatch Range and the High Plateaus divide the Great Basin from the country drained by the Colorado River. That vast region was in the Eocene covered by a lake perhaps as large as all the Canadian lakes combined. During Cretaceous and Eocene time the Wasatch and Uintahs were intermittently rising and being planed down by erosion nearly as fast as they rose. Major Powell holds that the upheaval of the Uintahs has suffered a degradation in areas of maximum erosion of no less than 30,000 feet, and Captain Dutton estimates the degradation of the Wasach at 50,000 feet. Sedimentary beds 6,000 to 15,000 feet in thickness were deposited over an area of 100,000 square miles. The whole series abounds in coal and carbonaceous shales and remains of land plants, and this indicates that throughout that inconceivable stretch of time the province as a whole remained almost on a level with the ocean and that therefore the sedimentary deposit must have sunk as fast as it was laid down. But faster than these great ranges were devastated they continued to rise. Query: Had the rising of the mountains and the sinking of the sedimentary deposit the relation of cause and effect, and what initiated the majestic movement? Captain Dutton is inclined to ascribe its initiation to that mysterous plutonic force whose operations never cease, and which constitutes the hardest problem of dynamical geology.

The Eocene lake dried up and disappeared at the close of the Eocene period. Then began the destruction and dissipation of these great bodies of sediment. Then, too, began the gradual elevation of the entire region, which has gone on, as well as the erosion, to the present time, and is not yet ended. Through the plexus of streams which unite in the Colorado River, the waste of the land has been carried to the Pacific. This river system is older than the structural features of the country. Since the river began to flow, mountains and plateaus have risen across its track and those of its tributaries. As the mountains and plateaus rose up, the streams cut their way through in the same old places. The course of the streams was determined by the configuration of the old Eocene lake bottom at the time the lake was drained. This was a period of slow uplifting, and of stupendous erosion as well. The swells in the bottom of the old lake became centers or axes from which erosion proceeded radially outward, dissolving away the strata in all directions. After the lapse of a long period, the newest or uppermost stratum encircled these centers of erosion at a great distance, the next group below encircled them a little nearer, and so on. North and west of the Colorado, five of these centers are easily discerned. Captain Dutton describes one of these — the San Rafael Swell, lying between the Green River and the Wasatch Plateau. Standing on the eastern verge of the Wasatch, at an elevation of 11,000 feet, and looking eastward, the eye sweeps a semi-circle with a radius of more than seventy miles. " It is a scene of desolation and decay; a land dead and rotten, with dissolution apparent all over its face. It consists of a series of terraces, all inclining upward towards the east, cut by a labyrinth of deep narrow gorges, and sprinkled with numberless buttes of strange form and sculpture." There are five of these concentric lines of cliffs. One after another in orderly succession, the whole stratagraphic series from the base of the Mesozoic to to the summit of the Eocene, 10,000 feet in thickness, have been stripped off, and the edges of the remaining portion form the successive cliffs which bound the encircling terraces — " cliffs of strange aspect, winding in all directions, until they sink below the horrizon, swing behind some loftier mass, or fade out in the distant haze. Wonderful, at times, is the sculpture of these majestic walls, and very striking the coloring, belts of fierce staring red, yellow and toned white, intensified by alternating bands of dark iron gray." Since the river began to flow, its sources have been rising and its slope increasing. The land is dissected by the river system, and so deep and elaborate are these cañons that it is impossible for unwinged creatures to traverse parts of the country at all.

The High Plateaus are the remnants of masses of Tertiary and Cretaceous strata left by the denudation of the country to the east and south. Until near the close of the Pliocene they were the scene of great volcanic activity; and the enormous floods of lava poured out upon them preserved them from degradation. In the Miocene the climate appears to have been humid, and to this period the greater part of the denudation is assigned. The Pliocene is supposed to have witnessed the raising of the Wasatch, the Uintahs, and the High Plateaus 10,000 to 12,000 feet, and of the adjoining Great Basin area 5,000 to 6,000 feet. That was the era of the vast longitudinal faulting, of the resulting volcanic activity, and probably of the gradual development of our present arid climate. On the west side, the high central zone of Utah is approached by the long slopes of great monoclinal flexures or the grand cliffs produced by displacement; on the east side, by cliffs of erosion.

LAKE BONNEVILLE.—Mr. M. E. Jones, in his "Salt Lake City," gives the following spirited picture of the immediate valley of Great Salt Lake, with which this topic may be appropriately ended:

"In recent geological times the lake covered most of western Utah; the mountain ranges were islands or peninsulas of great length; the water was fresh, 1,000 feet deep, and had an outlet through Red Rock Gap, where a large, deep, but gentle river went to swell the great Columbia in its onward march to the ocean. The grandeur of this great sea (18,000 square miles in extent), as large as Lake Huron, with its icebergs floating off from the glaciers which plowed their way down the cañons, its tremendous waves, its great rivers, all set in a border of dense black forests, and lit up by long chains of lofty mountains, glistening almost to their bases with perpetual snow, can hardly be described or imagined. The lake teemed with fish and fowl of almost every variety, and the country with deer, elk, buffalo, mountain sheep, bear, foxes, wolves, etc. The beautiful columbine, the pride of Utah, with its white, pink and lavender blossoms, grew on every hill; the blue foxglove and larkspur in every valley; the open country was carpeted with luxuriant grass. But the climate was gradually changing; there were periods when the snow crept far down the slopes and into the valleys, and periods when it receded far up the mountains; but at the end of every cycle it was found to be gradually disappearing; so the forests climbed the mountains, the valleys opened out into magnificent parks, covered with grass and decked with multitudes of beautiful flowers, and enriched with clumps of firs and scattered pines and groves of deciduous trees. Still greater changes came over the beautiful land. The volcanoes were still in active operation, hurling ashes and lava into the lake, belching out fire, and painting the sky with the ominous cypress tree cloud. The river, the outlet, gradually wore away its lime-stone bed to a depth of 360 feet, draining large areas, and now the increasing warmth of the climate parched the land, dried up the little streams and contracted the large rivers till the water of the Lake no longer flowed from the outlet, and the mineral matter in it increased with the evaporation till the fish all died. The snow disappeared from the lofty mountains, the forests faded away in the valleys, leaving only cottonwoods and willows sprinkled along the streams. The lake had dried up to one-tenth of its former size, being about seventy-five miles long by fifty wide, but the valleys were still covered with luxuriant grass, the home of much game and the most pleasant spot between the mountains, when the Indians came and settled here. How long they were here, no one knows, but it was a long time, sufficient for a widespread opinion to get out that somewhere in the great West there was a strange salt sea and fertile valleys held by powerful Indian tribes."

CLIMATE; METEOROLOGICAL; SANATORY.

TEMPERATURE AND PRECIPITATION. — The effect of residence in any country, upon the physical well-being of man, is mainly a question of climate. In a mountainous district like Utah, the climate will, of course, vary considerably with varying altitudes and exposures. The inhabited parts of the territory range, in general, between 4,300 and 6,300 feet in absolute altitude, but 70 per cent. of the population is settled in valleys not more than 4,500 feet higher than the sea, and 60 per cent. of them in the valleys of Great Salt Lake and Utah Lake. In this basin the air is dry, pure, elastic, transparent and bracing; and the temperature compares favorably, in respect of equability, with that of the United States at large, and especially with that of Colorado and Nevada, and the Territories north and south of Utah.

The annual range of temperature and amount of precipitation at Salt Lake City, may be seen in the following table, prepared from the records of the U. S. Signal Service at Salt Lake City:

TABLE NO. 1.—*Mean, extremes, and range of temperature, annually, and amount of precipitation, period extending from 1875 to 1892, inclusive.*

YEARS.	TEMPERATURE.				PRECIP'N.
	Mean.	Maximum.	Minimum.	Range.	Inches.
1875............	51.26	101	5	96	23.64
1876............	50.64	97	7	90	21.23
1877............	51.00	98	3	95	16.35
1878............	51.29	97	5	92	19.75
1879............	53.20	97	—10	107	13.11
18•0............	54.00	95	2	93	10.94
1881............	51.54	100	2	98	16.88
1882............	49.20	96	0	96	15.98
1883............	50.80	100	—20	120	14.24
1884............	50.90	93.2	—13	106.2	17.52
1885............	52.30	100.3	4.6	95.7	19.69
1886............	51.60	99.1	—2.9	102.0	18.89
1887............	52.70	97.9	8.7	89.2	11.66
1888............	53.00	98.2	—16.7	114.9	13.62
1889............	52.70	102	5	97	18.46
1890............	51.30	100	—6	106	10.33
1891............	50.60	98	0	98	15.92
1892............	52.70	100	1	101	18.35
Average........	51.2	98.3	1.5	99.8	16.20

The annual mean of Salt Lake City places it very near the isothermal line of 50°, which crosses nearly 15° of latitude on each continent, owing to the disturbing influences of oceans, winds, and elevations; starting on Puget Sound and passing through or near Salt Lake City, Sante Fé, Denver, Burlington, Pittsburg, New Haven, Dublin, Brussels, Vienna and Pekin. The summer and winter means describe the same undulations in traversing the continents, and they are more indicative of the climate in its relations to animal and vegetable life than the annual mean. The mean annual temperature of New York and Liverpool are the same, yet throughout England the heat of summer is insufficient to ripen Indian Corn, while the Ivy, which grows luxuriantly in England, can hardly survive the severe winters of New York.

The following table, prepared from the records of the Signal Service, exhibits the average of the extreme range of temperature, not of the maximum and minimum of each day in the month, but of the single highest and lowest reading in each month, and also of that day in each month on which the range was greatest; the mean temperature by months; the mean daily range; direction and velocity and total movement of the winds; relative humidity, precipitation, etc., by months, period of observation extending from 1875 to 1891 inclusive. The figures under the head of "average cloudiness," give the total actual cloudiness from daily observations. For the purpose of observation the day is divided into tenths, the cloudy tenths noted, added together and divided by the number of days in the month. Thus the figures represent not 50.1 full cloudy days in the year, but the equivalent of 50.1 full cloudy days distributed throughout the year.

TABLE No. 2.—*Averages of each month for seventeen years, of seasons and years—1875 to 1891, inclusive; also the average for 1892.*

YEAR.	TEMPERATURE.						WIND.				PREC'T'N.		
	Monthly mean.	Mean of highest in each month.	Mean of lowest in each month.	Mean of extreme monthly range.	Mean of extreme daily range.	Mean daily range	Prevailing direction.	Average hourly velocity.	Total movement.		Mean relative humidity.	Rain fall, Inches.	Average cloud-ness. [0.10]
December.............	33.3	51.1	13.6	44.7	25.0	14.6	S E	4.1	3,032		65.9	1.73	5.7
January..	28.0	17.2	10.5	48.1	24.3	15.7	S E	4.1	3,155		63.8	1.48	5.1
February.............	33.6	53.1	13.6	46.1	26.5	15.7	S E	4.6	3,163		61.7	1.35	5.2
March............	41.9	65.5	23.3	47.8	32.6	18.5	N W	5.5	4,151		53.7	2.00	4.9
April	49.9	72.5	31.0	47.0	35.2	20.1	N W	6.1	4,435		49.7	2.39	5.0
May	58.1	82.7	37.4	50.0	38.8	22.2	N W	6.3	4,647		45.7	1.66	4.3
June	67.4	90.9	45.2	49.9	40.4	24.4	N W	6.2	4,425		39.5	0.77	3.1
July....................	75.7	95.4	53.6	45.9	41.5	24 8	N W	5.6	4,131		37.4	0.40	3.2
August	75.0	95.0	52.5	46.7	40.4	24.1	S E	5.5	4,418		37.5	0.75	3.1
September............	64.6	87.1	43.3	49.5	39.5	23.6	N W	5.4	3,846		38.2	0.88	2.8
October...............	51.9	75.6	30.8	49.6	35.0	20.5	N W	4.9	3,630		47.6	1.67	3.7
November........ ...	39.2	61.6	18.6	46.2	30.0	17.5	N W	3.9	2,815		56.3	1.36	4.4
Winter.............	31.6	50.6	12 6	46.3	25.3	15.3	S E	4.3	9,350		63.8	4.56	16.0
Fall........	51.9	73.6	30.9	34.8	34.8	20.5	N W	4.7	10,291		47.4	3.91	10.9
Summer	72.7	93.8	50.4	47.5	40.8	24.4	N W	5.8	12,674		38.1	1.92	9.4
Spring.............	50.0	74.8	30.6	48.3	35.5	20.3	N W	6.0	13,233		49.7	6.05	11.2
Year	51.5	73.1	31.1	47.2	34.1	20.1	NNW	5.6	45,548		49.7	16.44	50.5
Average for year 1892	51.2	61.6	31.2	48.5	32.1	19.8	N W	5.3	44,976		56.8	14.08	4.8

The slight discrepancy in precipitation between tables one and two is due to the dropping of hundredths in averaging the months. In tropical regions there is but slight difference between the mean of the hottest and the coldest months. At Singapore the difference is 3½°, at points in Siberia it is 100°, at Quebec it is 60°, at Salt Lake City it is 46.3°. A summer mean of 72.5° may seem high, but to the denizen of Salt Lake Valley the mountains are convenient, and the dry and absorbent nature of the air and the influence of Great Salt Lake constitute a modifying cause, making either extreme heat or cold less oppressive by perhaps 10° than the actual reading indicates. The relative humidity is 48.9 per cent. At Philadelphia it is 73 per cent, at Denver 49.7 per cent, at Los Angeles 66.2 per cent, at Colorado Springs 55.5. For spring, summer, and fall it is 45.1 per cent, while for summer it is 38.1 per cent. The rainfall is 16.44 inches a year. In the humid region, same latitude, it is 40 inches in a year. Fort Laramie, Sacramento, Santa Fé and Denver have about the same as Salt Lake City, while in general east of the 100th meridian it is 40 inches, 60 per cent of which at once goes off in the surface drainage. The days on which there is precipitation average one in four, but severe storms are very rare, and the days when the sun is not seen are extremely few. The mean air pressure is 25.63 inches. Water boils at 204.3°. The prevailing winds are from the north-northwest, and the most windy months are April, May, and June. The mean velocity of the winds is 5.6 miles per hour, and the average annual total movement is 45,548 miles. At Philadelphia it is 100,000; on the ocean it is 150,000. There are no cyclones, and severe lightning and thunder are very infrequent.

The climate attains that medium between the rigor of the great fresh water lakes region, and the eternal summer of Florida and Southern California, which makes it both healthy and agreeable. The normal winter has thirty to forty days of moderately cold weather, with enough snow for a week or two of sleighing. Indian summer holds on to Thanksgiving, while the planting season begins in February. There is comparative exemption from the changeable weather and raw winds of spring in the North and East. Only in one month out of five does the range in temperature exceed 50°. The sun shines perpetually, the air is invigorating, the rapid radiation assures cool nights. But no words or meteorological statistics can convey an adequate idea of the charm of the climate, which continues to grow upon one no matter how long a resident.

Hardly any form of disease originates in Utah, while upon many diseases contracted elsewhere simple residence and use of the thermal waters in the city, and of Great Salt Lake in the bathing season, are more beneficial than ordinary medical treatment. There is no malaria, asthma is impossible; the Utah Hot Springs north of Ogden are a specific for rheumatism; pyæmia from surgical operations is exceedingly rare; pulmonary complaints are stayed in their ravages if not cured; there are none of the more virulent fevers, and diphtheria takes on a relatively mild type.

People from boasted sanitariums are constantly dropping into Salt Lake City and Valley, experiencing relief, often unexpectedly, settling down and growing robust. Every Utah reader of these lines will readily call to mind examples of this within his personal knowledge. The Territory is full of octogenarians enjoying a serene and hearty old age, who will be superseded by centenarians when the natives have had time to become such.

SALT LAKE CITY A PERFECT HIGH ALTITUDE RESORT.—Dr. H. D. Niles, M. D., of Salt Lake City, prepared the following:

"The sanitary advantages of Salt Lake City are,—

"1. A distinctly local climate, apparently possessing in a marked degree the popular requirements of a high altitude resort.

"2. Unexcelled salt water bathing, the peculiarities of which may indicate unusual remedial virtues.

"3. Hot, warm, and cold sulphur springs, of alleged marvelous curative properties.

"A high altitude resort should possess the greatest possible dryness and equability, an elevation of from 3,500 to 8,000 feet, the greatest number of sunshiny days during the year in which the invalid may enjoy outdoor life, comparative freedom from wind and sand-storms, a proper temperature, and certain other qualifications not of a purely climatological character.

"DRYNESS.—First in importance is dryness of the air as indicated inversely by the relative humidity. Salt Lake City's air averages 49.7 per cent relative humidity, and consequently lacks 50.3 per cent of saturation. El Paso and Santa Fé are the only places in the United States where observations have been taken at which the air has a greater absorptive capacity than at Salt Lake City.

"Recent investigations have strengthened the prevailing opinion of the great value of the absorptive and aseptic qualities peculiar to dry air in the treatment of pulmonary diseases. It is well known that warmth and moisture favor decomposition and the generation of micro-organisms and lessen the vapor transpiration. Dry air, on the contrary, destroys or retards germ-life, in and out of the lungs, and increases the amount exhaled. The moisture thus exhaled may serve as a vehicle for the removal of effete matter, wasted tissue, and germs of disease.

"Dr. Denison, whose original researches have added much to our knowledge of high altitude climates, suggests an ingenious method for calculating the excess of moisture exhaled in a cool dry air over that in a warm moist air. In this calculation Glaisher's estimate of the weight of vapor in grains in air of a given temperature is adopted, and it is assumed that the dew-point in exhaled air is 94° F., and that it is saturated. Apply this method to the air of Salt Lake City and Los Angeles, in the case of a good sized man, who, we will suppose, breathes twenty times a minute and thirty inches per breath on an average, and we find Salt Lake's excess in transpiration over Los Angeles to be about four ounces daily. The remarkable aseptic and absorptive properties of the air of Salt Lake City unquestionably have a favorable influence in cases of surgery.

"ATMOSPHERIC PRESSURE.—In regard to altitude, Salt Lake City is located immediately at the foot of a range of mountains, and the health seeker is thereby enabled to select any altitude from 4,300 to 8,000 feet above sea level, as may be best suited to his particular case; or he may vary it as the progress of the case demands, and still be near enough to the city and to Great Salt Lake to enjoy all their advantages. In rare cases where still higher altitude is desirable, it may be had, but at a greater distance from the city.

"FAIR DAYS.—Salt Lake City has an average of 277 fair days in the year; of the remaining 88 days, there are very few in which the sun does not shine a part of the day. The records of the Signal Service office for fourteen years shows that we have the equivalent of 50.5 full cloudy days in the year, and no more; about one day in a week.

"WIND.—The total movement of the winds is 45,548 miles per year; average hourly velocity, 5.6 miles; prevailing direction from the north-north-west. There is comparative freedom from high winds, and an entire absence of cyclones and hurricanes. High winds, like variability, have been regarded as necessary evils of high and dry places; in other words, it is the received opinion that places sufficiently sheltered by mountains to be protected from the winds cannot be very dry. Salt Lake City may be the exception that proves the rule; the city is certainly an exception if such is the rule.

"TEMPERATURE.—With an average annual mean of 51.5°, an average monthly range of 47.2°, and an average daily range of 20.1°, Salt Lake City has an exceptionally cool and equable climate. Possessing the advantages of seasonal changes in temperature, these changes are so gradual and the air is so dry that neither the cold of winter nor the heat of summer produces the unpleasant effects which they otherwise might. On the slopes and in the cañons of the adjacent mountains, in close proximity and convenient to the city, are places where the invalid may find both altitude and coolness during July and August, should an average of 75.6° be found too warm. From 1863 to 1888, inclusive, there were but twenty-one days on which the thermometer read below zero, and but four years in which it rose above 100° F.

"A very rare, if not unique, feature in the climate of Salt Lake City is its equability, as shown by a mean daily range of about 20°. Equability has been regarded as belonging exclusively to low and humid regions, and 'variability as a distinguishing attribute of all high and very dry places.'

"Mr. Brice says: 'It becomes a matter for careful study to determine wherein lies the happy mean between dry climates with great daily range, and moist climates with small daily range of temperature.' Other authorities are equally conclusive that equability is compatible with great dryness. The reply to the climato-therapeutist who has demanded dryness and equability combined, has been, 'You ask that which is impossible.'

"But it is not impossible, since Salt Lake City possesses this combination, as the Signal Service records of seventeen years prove. For spring, autumn and winter, when great daily range is most harmful, the mean is 18.70. It is greatest, to-wit, 24.4 in summer, when the temperature is high at the lowest, and when great diurnal range is agreeable and beneficial, rather than unpleasant and harmful. The hot months are, without doubt, the healthiest months. The mountain protection on the east and the presence of 2.000 square miles of salt water on the northwest, undoubtedly have much to do in producing this distinctly local climate.

"DIATHERMANCY AND OZONE.—It is generally believed that diathermancy of the air increases with the altitude, and that ozone is most abundant in sea-air and the air of mountainous regions. The air of Salt Lake City is a mingling of sea and mountain air. To sum up, it appears that in point of dryness of the air, equability of temperature, comparative exemption from high winds and sand storms, choice of altitude convenient of access from the city, between 4.300 and 8,000 feet, relative number of sunshiny days in which a patient may safely enjoy out-door life, and purity and stimulating quality of the atmosphere, Salt Lake City possesses the nearest approach to an ideal high altitude resort at present known.

"OTHER QUALIFICATIONS.—The location of the city and its soil are favorable for perfect drainage. The city drinking water is from the mountains, comes out of limestone, and contains a very small amount of organic matter. The soil of the adjacent regions produces in abundance and perfect in quality, all the fruits, grains and vegetables common to the latitude. The fruits are confessedly more fair and more highly flavored than California fruits, and the wheat and potatoes are unexcelled in quality. The markets are well supplied with fresh fish (trout),* and with different kinds of game in their season. Also the home supply is supplemented by that of California. The almost perpetual sunshine, the ever novel scenery, the fine drives, the trees, the lawns, the shade, and the softness of the air, draw the people much into the open air.

"GREAT SALT LAKE BATHING.—The percentage of solid matter in the water is about 17 as against about 2 per cent in ordinary sea water. The water is free from odor. Possibly the quality and large proportion of saline matter may have a sanatory significance. The only body of water having so high a specific gravity, namely, 1,110, is the Dead Sea.

"The calmness of the surface, the buoyancy of the water, its warmth, and the fine beach, combine to render bathing and swimming in the lake easy and pleasant exercise for the average invalid, and free from many of the dangers and difficulties attending ordinary surf-bathing. No living thing inhabits the waters of the lake, an advantage that will be appreciated by the sensitive and timid.

"Spacious and commodious buildings have been erected at the different resorts, and every effort is made to accommodate the rapidly increasing number of bathers; so that, the peculiarities mentioned excepted, these bathing resorts do not materially differ from the first-class ocean resorts.

"SULPHUR SPRINGS.—Hot sulphur springs boil out in great volume within the city limits. A company pipes the waters into a swimming pool and bathing house in the heart of the city. The water contains a small quantity of free carbonic acid and a large amount of sulphuretted hydrogen

---

* Lakes and private ponds swarm with German carp, and several of the former and some of the larger streams have been well stocked with rock bass, catfish, eels, etc., which are growing and multiplying with gratifying rapidity.

gas. These springs, and the natatorium in the city supplied from them, are largely patronized, and marvelous stories are told of their efficacy in rheumatic, syphilitic and skin diseases.* The water of Great Salt Lake is also piped into the city and run into an immense swimming pool.†

"As a health-resort, however, it is only candid to say that Salt Lake City has some drawbacks. There are no regular resort hotels. The streets are not paved, and the sanitary condition of the city is not what it should be.‡

"Sewerage is now being put in. The pavements and the resort hotels will soon follow. Street cars will soon run by electricity on thirty miles of the streets (they are already running on several routes), good lighting will accompany this improvement, plenty of water will be provided, dust and mud will be banished, and in the near future the health-seeker will find in Salt Lake City not only everything that contributes to the health and comfort of the invalid, but a most beautiful city in which to live."§

A NATURAL SANITARIUM.—The following letter, written from St. George, Utah, under date of December 30, 1888, by Mr. George H. Wyman, is so full of suggestions and experiences of value to the intelligent searcher after the "right spot," as to be of special value in this connection:

"It has seemed to me that it might be sufficiently useful to my countrymen, especially of the Eastern States, to warrant my giving my experience of Utah as having a climate favorable to the restoration of invalids. It is only for those having lung difficulties, and asthma in particular, these remarks are intended. For the last twenty years my search has been pretty constant to find the most favored region to live in, where a cure, or at least comfort, could be obtained. At first, because convenient, the Atlantic seaboard was tried, and at many of the most noted watering-places. It did not. Then I tried the Lake Superior region and the pine woods of Michigan, and two or three trips to the Adirondacks. After that, for a couple of months in summer, I was in Denver and the hills west of the Snowy Range. This was much better, and for three or four years I did well, but the seasons of out-door life were not long enough. The cold, rough winds came early. On the Pacific Coast I tried Santa Barbara, Los Angeles, and San Diego, with the adjacent mountains. The mountain climate was good, but in the fall and winter the constant rains and fog rendered that region unsatisfactory. I tried Arizona and New Mexico. The objection to the latter Territory in the winter time was its severe winds. Its high table-lands, and lying just south of the Rockies, brought a strong wind, making out-door life uncomfortable. In many places named on the Pacific Coast I found scores of invalids from the Eastern States, hoping to profit something from their stay through the winter. It became certain to my observation that only when they could rough it

---

* They are known to be almost a specific in kidney troubles, if not too complicated.

† The waters have since been piped and are now used in a natatorium in the heart of the city. In addition to this, the finest water for culinary purposes is brought in through an underground conduit from Parley's Cañon, ten miles distant.

‡ All this is now completely changed, as the writer would readily admit if here.

§ Salt Lake City has now only electric cars, running on more than forty miles of track inside the city proper; also an excellent sewerage system in full operation.

in a clear, dry atmosphere did they make any headway against disease. And so it seems to me that wherever such class of invalids can, both summer and winter, find most suitable for outdoor exercise and pastime, that place is best for them. This is the result of my observation and experience. For the last ten years it has been my habit to spend the spring and summer months at Salt Lake City, in Utah, and in the adjacent mountains, Parley's Park, Pleasant Valley, Ophir, and the Cottonwoods, going into the hills as early in summer, and as high up the mountains as I found convenient, and in the fall I came to St. George, in the extreme southern part of Utah, for winter. I have been best suited with St. George for the simple reason that I find it sufficiently warm, and with less storms of wind, rain, or snow than any place I know. So that one can drive out here with more sunshine to cheer him during the winter days than in any place I have mentioned, and with less annoyance from wind, dust, or rain. St. George is a quiet town, of perhaps 1,000 inhabitants, and bordering on the Virgin and Santa Clara rivers at their junction.

"The soil is sandy and produces, besides the ordinary field crops, cotton, a large variety of fruits, and the valley is noted for its fine grapes and excellent wine. To get here from Salt Lake, after reaching the terminus of the railroad at Milford, there are about 100 miles to go by wagon. An invalid unused to our ways of travel might regard this journey as serious. If obliged to come by stage, and without a rest, it might be so, but if he comes with his own team, and a proper camping outfit, he is just beginning to get well, and can pursue his cure at his convenience all the way here, and be much stronger and happier because of his journey. There will be some mountains to cross, 7,000 or 8,000 feet above sea level, but with good roads. Half the distance over, the traveler begins to descend into this valley, the upper part of which is liberally covered wi h black basaltic rocks. At St. George there are large, pleasant fields free from rocks, and well cultivated. My way of getting here has always been to drive all the way from Salt Lake. One needs a carriage of his own here, and it is desirable to come with his team if he can. For pastime a gentleman could buy a lot and indulge in grape culture. If he likes shooting, there are wild geese, ducks, and quail here in winter. All the needful things to live on are to be had, and good milk, honey, and wine can be bought at almost every house. About the first of May it is generally desirable to leave here for the north. I am satisfied there are many invalids who would be more likely to recover health here than in any of the places I have mentioned, but my recommendation is to those only whose condition will admit of camping out, and an out-door life generally."

EFFECT ON CONSUMPTION.—The following pathetic appeal was received by Prof. M. E. Jones, of Salt Lake City, from B. L. Bonsall, of Delair, N. J., under date of August 10, 1889:

"*Dear Sir:*—'We' received a copy of Marcus E. Jones's pamphlet through your courtesy.

"As one of the editorial 'we,' I am a doomed man if I remain here, and may be anyhow, no matter where I go.

"But, having the means to try other climates, I must confess your pamphlet has influenced me greatly.

"What causes me to hesitate most about starting for Salt Lake this month is the fact that our doctors all agree that to go west or northwest to spend winter would be suicidal, because of severity of season, etc., being as bad as our own. Go southwest they say, or south, but Utah is not southwest, but as far north as Colorado, which is claimed to be all right for summer, but a poor place for winter.

"Physicians say I am not in the last stages, and although I had three hemorrhages last Sunday, I have been favored with strength to be about my room again, and will make the trip to Utah *alone*, if I come at all. *Now, shall I come*, or would you advise a warmer place? *Are* your winters only six weeks long and are they mild enough for a man with lung trouble?"

And this is Mr. Jones's reply, addressed to *The Salt Lake Tribune:*

"*Editor Tribune:*—The subjoined letter is a sample of what I frequently receive from various parts of the country. Some lives may be saved by giving this to the public and by answering its questions.

"I did not send the gentleman a copy of the pamphlet, nor do I know who did.

"Before answering this letter I consulted two of our leading physicians, to be sure that I made no error. They say emphatically that 'this is just the time to come,' for from now till Christmas the weather is liable to be perfectly uniform, no sudden changes, but getting gradually cooler day by day and becoming more and more bracing. From Christmas till the last of January is our brief winter, which is more like the flurries of snow in November than a real winter. The temperature may go to zero for one night, or even some lower, but the air is so dry that the cold will not be felt as much as the damp, chilling air of New Jersey, 20 to 30 degrees higher. Our winds are slight and we have less sudden changes by fifty per cent than any place on the Pacific Coast.

"It is true that we are about on the same parallel as Northern Colorado; it is also true that Alaska is north of or on the same parallel as Labrador, but Sitka, Alaska, has the same climate as Virginia. The climate of Salt Lake is milder than that of Santa Fé, New Mexico. We are free from the terrific winds and chilling storms that sweep down from the Rockies. It is not often that we have a week's sleighing in the whole winter. We are in a beautiful valley protected on all sides but the south, from which comes warm breezes in winter. Colorado may be all right in summer (I doubt it), but Salt Lake can be depended upon all the year around.

"The doctors say that the fact that Mr. Bonsall had a hemorrhage, is all the more reason why he should come at once to avoid a recurrence of them.

"If he is in the last stages he will probably die on the way; but if he is not, I do not hesitate to advise him to come without delay. A warmer climate will only debilitate instead of invigorate him.

"Yes, our winters are only six weeks long on an average. I have picked wild flowers on the bench back of the city on February 15 or thereabouts nearly every year since I came here, and last winter I found one in bloom on January 1. It was a hold over. The ground was slightly frozen at the time and it was too cold for new flowers to come out.

"Our winters are certainly mild enough for people with lung troubles, and not only that, they are very beneficial.

"The low altitude and great dryness of Salt Lake place us at the head of all pulmonary sanitariums.

"Let the consumptives come and see for themselves. They will die if they stay at home. If this does not help them, they might as well settle here and live while they can, for I know of no other place that will help them."

FARMING; STOCK-RAISING.

AGRICULTURAL COLLEGE.—An agricultural college and station has been established at Logan, the capital of Cache County. The board of trustees as now organized includes William S. McCornick, Salt Lake City, president; William N. Brown, Provo; Christian F. Olsen, Hyrum; Robert W. Cross, Ogden; Melvin B. Sowies, Salt Lake City; John E. Hills, Provo, James T. Hammond, Logan. John T. Caine, Jr., is secretary of the board, and H. E. Hatch, treasurer. The faculty are: Jeremiah W. Sanborn, B. S., president, professor of agriculture; Evert S. Richman, M. A. S., professor of horticulture and botany; William P. Cutter, B. S., professor of chemistry; Abbie L. Marlatt, B. S., professor of domestic economy; J. M. Sholl, professor of mechanical engineering; Alonzo A. Mills, B. S., farm superintendent. The station staff is organized as follows: J. W. Sanborn, B. S., director; E. S. Richman, M. A. S., horticulturist and entomologist: W. P. Cutter, B. S., chemist; A. A. Mills, superintendent of experiment work; J. R. Walker, clerk and stenographer; H. E. Hatch, treasurer. A farmhouse, a laboratory, a barn, and two cottages are being built at a cost of over $14,000, the legislature of the Territory having aided the station with liberal appropriations for buildings, live stock, etc. Eighty-five acres of ground are devoted to experimental purposes. The Territorial Legislature, now in session (January 1892), has before it a bill for a liberal appropriation to aid the college, and still further extend its field of operations and increase its usefulness.

IRRIGATION.—The area of irrigable-arable land in Utah has been given under a previous head. Although crops are grown in favorable spots without irrigation, yet irrigation is indispensable to the Utah farmer. With the canals and acequias made, watering costs at the outside $3.50 per acre; and it enriches the land and assures a full crop. If water can be so distributed as not to run over and off from the land, it is held that it will impart more of the elements of plant growth than the harvest extracts. A certain forty square miles in Valencia, Spain, under the canals of the Turia, sustain 70,960 souls, besides the population of the city of Valencia. At one-fourth of this density, Utah's irrigable-arable lands will sustain 1,250,000 souls. Civilization is indigenous only in rainless countries, where man controls seed-time and harvest. Any farmer in the world might well choose to do his own watering (if he could), rather than be at the mercy of the capricious skies which bend above the so-called humid regions.'

Twelve years ago there were 10,000 miles of acequias, large and small, watering as many small farms in the valleys of Utah. No doubt the farms and the miles of irrigating channels have both increased in number fully 100 per cent since that time. The necessity for irrigation tends to keep down the size of farms, and this tends to high cultivation. With a strong and fertile soil, an unclouded sky, a clear atmosphere, an equable climate, reliable seasons, plenty of water, and a multitude of husbandmen relatively to the acres cropped, if our product per acre does not double that of the older States we have nobody but ourselves to blame. Nature has done her part. Our cereals, fruits, and vegetables are of superior quality, and many things which never grew under Utah skies are in Eastern markets ticketed as Utah productions.

CROPS.—The report of the Department of Agriculture, 1888, pp. 431-32, is the authority for the following table:—

PRODUCT.	Average yield per acre, whole country.	Average yield per acre in Utah.	Average price per unit, whole country.	Average price per unit, in Utah.
Corn, bushels......	20.1	21.6	$ .444	$ .75
Wheat.............	12.1	19.0	.681	.61
Rye	10.1	8.3	.545	.47
Oats................	25.4	26.5	.304	.43
Barley.............	19.6	22.2	.519	.58
Potatoes...........	56.9	90.0	.682	.36
Hay, tons	1.10	1.20	9.97	6.90

A Bulletin of the Agricultural Department, published in December, 1890, has the following table:—

Average farm price of Agricultural Products, December 1, 1890.:

	Corn, per bushel.	Wheat, per bushel.	Rye, per bushel.	Oats, per bushel.	Barley, per bushel	Buckwheat, per bushel.	Potatoes, per bu.	Hay, per ton.
General average	.50½	.84	.63	.42	.65	.58	.78	$7.74
Utah average	.63	.78	.63	.55	.75	.72	.75	8.00

The same report (of 1888) shows the average cash value per acre of farm products in the whole country for 1887, and for Utah as follows:

PRODUCT.	Whole Country.	Utah.
Corn	$ 8.93	$16.20
Wheat	8.25	11.50
Rye	5.49	3.90
Oats	7.74	11.40
Barley	10.15	12.87
Potatoes	38.82	32.41
Hay	10,98	8.28
Total	$90.36	$96.05

The result is seven per cent. in favor of the Utah farmer, and in this is necessarily reckoned considerable dry farming. We place no great reliance on these figures, but they are probably as accurate for Utah as for other parts of the country. Yet any one who lives in Utah and sees the third heavy cutting of alfalfa per season, will have his own opinion about the yield of 1.2 tons per acre. It is safe to say that three-fourths of the land mown in Utah is alfalfa, and the average yield is nearer eight tons per acre than 1.2 tons.

In another table, in the same report, Utah is credited as follows: to-wit:

PRODUCT.	Crop 1887.	Av. yield per acre.	Acres in each crop.	Value per unit.	Total Value.
Corn, bushels	285,000	21.6	13,197	$0.75	$ 213,750
Wheat, "	1,971,000	19.0	103,738	.61	1,202,310
Rye, "	19,000	8.3	2,287	.47	8,930
Oats, "	786,000	26.5	24,658	.43	237,980
Barley, "	660,000	32.2	29,750	.58	342,800
Potatoes, "	1,088,000	90.0	12,084	.36	391,680
Hay, tons	194,762	1.2	162,302	6.90	1,343,858
Total			353,016		$3,881,308

AREA. — In an Agricultural Department Crop Bulletin, published in December, 1890, the following, concerning Utah crops of 1890, occurs:

PRODUCT.	Acres.	Bushels.	Value.
Corn	35,175	739,000	$ 502,299
Wheat	130,251	2,279,000	1,777,927
Oats	38,491	1,059,000	582,177

This shows an increase over 1877, in acres, of forty per cent.; in bushels, of thirty-four per cent; in value. of seventy-three per cent.

The table, it will be seen, includes corn, wheat, rye, oats, barley, potatoes and hay only, and exhibits a very gratifying enlargement of farming operations in three years. Adding the area in all other crops, inclusive of gardens and orchards, beets and sugar cane, we should have fully half a million acres planted, and realize from it all not far from $10,000,000. In acreage cultivated, Utah exceeds either Wyoming, New Mexico, Idaho, Arizona, Nevada, Delaware, Rhode Island, Montana, or Colorado.

Yield and Quality of Products.—The amount and value of Utah's principal products for 1891 are as follows:

Product.	Acres.	Bushels.	Value.
Corn (72c. per bushel)	8,776	165,067	$ 118,848 24
Wheat (80c. ")	110,114	2,409,454	1,927,563 20
Oats (47c. ")	32,763	1,132,218	532,142 46
Rye (61c. ")	3,759	45,204	27,574 44
Barley (50c. ")	7,358	212,546	106,273 00
Potatoes (53c. ")	7,845	935,874	496,013 22
Hay ($13.58 per ton)	80,647	(tons)120,572	1,637,367 76

Total..$5,845,782 32

Wheat.—Fair Utah wheat ranks in the East with the best No. 2 red, which is the highest grade that appears in most of the Eastern markets. The choicest varieties are a unique product with scarcely an equal in America. Utah wheat has a brighter, larger kernel than that of the East, and, though no handsomer than that of California, it is firmer, and its nutriment more concentrated. As high as sixty bushels per acre have been raised here, but the average yield is not more than thirty.

The largest wheat regions are in Cashe and Utah Counties, closely followed by San Pete, Salt Lake and Weber Counties. The great staple is, however, raised throughout the entire Territory. Utah wheat rarely falls in price below one cent per pound, free on board.

Oats.—Parties who are keeping up work-horses pay 25 to 30 per teen more for Utah oats of ordinary quality than for a fair grade of Eastern. Utah oats have ranged in price during recent years from 1¼ to 1¾ cents per pound on cars. Large farms have been known to realize an average of eighty-five bushels to the acre by high cultivation.

Barley.—Usually Utah barley is of magnificent appearance. In recent years, brewing barley has been exported to St. Louis, Milwaukee, California and other points, where it grades up to the best Canadian brewing. It is the use of this barley that gives Utah beer so high a standard. Indeed, the White-club brewing barley will hold its own anywhere as a strictly fancy product.

Rye.—There are a few cars of rye annually offered at figures over one cent per pound. The quality is superb and the yield fair.

Corn.—Utah does not rank as a corn country, and rarely has any for export. The hot, sultry nights which corn requires are not characteristic of the climate. Still, it must not be supposed that farmers cannot raise corn in Utah, as more than 500,000 bushels are annually produced.

Grasses.—In the improvement of Utah lands, there remains untilled, and scarcely prized, a considerable area of rough ground too dry for grass and too broken and stony for grain. This is now being utilized for alfalfa, which succeeds almost anywhere in Utah. All the other tame grasses common to this latitude do well in Utah.

POTATOES.—The Utah potato has a reputation for excellence all over America, and even in Great Britain. For many years great quantities have been exported from the Territory.

The bulk of the potatoes raised and marketed are Early Rose, Early Goodrich, Willard and the Peerless. The King of the Early, Peerless and Compton's Surprise, yield, in favored localities, about 400 bushels to the acre. With high cultivation 800 bushels have been raised to the acre. With proper treatment potato-growing does not impoverish the land, some of the best results coming from ground that has been in potatoes for the last twenty years.

OTHER ROOTS.—Utah has a good reputation for carrots, which sometimes yield, of good quality, as much as 1,000 bushels to the acre; also, for tomatoes, onions, turnips, parsnips, radishes, etc. Beets thrive well; large tracts of low-lying lands on the western side of the Salt Lake Valley, and elsewhere, are now planted in sugar beets for the manufacture of sugar. These find a ready market at profitable figures at the Lehi Sugar Factory, which now supplies a great percentage of Utah's sugar. (See next page.)

GREEN STUFFS.—Of green stuffs Utah annually exports considerable quantities of cabbages, cauliflowers, melons, squashes and celery—the latter growing exceptionally fine.

HOPS are also native to Utah, their trailing vines overrunning every other kind of foliage in many of the cañons. One or two parties, notably on the Provo Bench, have heeded this suggestion, and are now growing a superior quality of hops.

The net value of these garden products in 1889 has been carefully estimated at $2,550,000, and in 1890 at considerably more, and a further increase for 1891 is noted.

ALFALFA.—The greatest farm crop of Utah is alfalfa. On good soil with plenty of water it is cut three or four times in the season, the total yield approaching eight tons per acre. The third growth is usually allowed to seed. After threshing, it is still pretty good fodder. It is held that the ground is benefited by its growth. The roots go deep for moisture and sustenance, and there is a perpetual rain of the leaves upon the soil. It is not adapted to other than dry climates. Steady sunny weather is required to cure the heavy growth. It brings the grower $5 to $10 per ton, and he gets 5 or 6 cents per pound for the seed; 400 pounds of seed is a heavy yield per acre, yet not uncommon. Both hay and seed find a market in adjoining Territories. It is better than anything but corn for fatting steers for market, unless the grain is ground and mixed. A steer consuming 33 pounds per day will put on one pound additional weight, thus by the animal mechanism transforming a ton of alfalfa into thirty pounds of fat beef. It is a beauty forever in the landscape; once well seeded there is nothing to do but water and cut and make hay of it and feed it. Every year in Utah its culture widens. It is far the best crop the Utah farmer can grow, potatoes possibly excepted, and it is the easiest grown. Among the crops it is what the labor-saving machinery is to human muscle. It is an untold treasure in Utah and in all the arid region. He must be a poor farmer who cannot make $40 an acre per year out of alfalfa.

SUGAR BEETS.—Investigation having shown that Utah, with a soil and climate perfectly adapted to the growth of sugar beets, was yearly paying abroad about one million dollars for sugar, a company, called the Utah Sugar Company, was recently organized with ample capital, the shares subscribed, the subscriptions called in, and a plant costing, altogether, $400,000, and capable of crushing daily 350 tons of beets and turning out 40 tons of refined sugar, has been put in near Lehi, Utah County, thirty miles south of Salt Lake, and for the greater part of two years has been running to nearly its full capacity. The sugar is the finest grade of granulated, has a slightly reddish tinge, and its sweetening qualities are equal to the best. It is now to be found in nearly every grocery store and on nearly every table in Utah.

The Sugar Company issued a circular urging farmers to grow sugar beets, and offering to pay $4.50 per short ton of beets delivered at the factory. This is the highest price paid for beets in any part of the world. A man who understands himself can grow 30 tons per acre on good ground having sufficient water, and at a cost not greatly exceeding that of an acre of corn. This industry opens a new and considerable source of revenue to the already fortunate farmer of Salt Lake Valley and other parts of Utah. It will take, perhaps, 3,000 acres to feed this mill, and Utah consumes the utmost sugar output of three or four such mills.

BEAR RIVER VALLEY LANDS.—The big irrigating enterprise of the Bear River Canal Company has been mentioned before, and also the appropriation of Bear Lake as a reservoir by the company. The latter step forever assures the supply. The headworks for the canals are in Bear River Cañon. A canal of 1,000 second-cubic feet capacity is taken out on each side of the river. That on the left bank is carried down along the base of the mountains about forty miles to Utah Hot Springs. In this vicinity a branch is led off toward the lake, where, around Plain City, there is a large body of warm, sandy, rich land. The main canal goes on to Ogden. There, exchanged with the users of Weber River water, the latter is to be taken out high up in Weber Cañon and carried out upon the sand ridge south of Ogden.

The canal on the right bank, when it reaches the valley-plain, strikes up the valley diagonally three miles to near the Toponce Ranch, where it is carried over the Malad, here 100 feet below its banks, on an iron viaduct costing $30,000. Thence it is led around the northeastern edge of the valley past Point Lookout and the Walker Ranch toward Blue Creek, about forty miles. Soon after reaching the plateau it throws off one branch which goes down near Bear River past Corinne to the lake, about 30 miles.

It throws off a second branch west of the Malad, which runs southward to near Little Mouutain, and then westward to the main canal.

The Canal Company offer these lands for sale at $25 to $35 an acre, which includes $10 for perpetual right to one cubic foot of water per second for each 80 acres. The yearly rental or maintainance tax it is proposed to put at $1.50 or $2.00 an acre of land watered. The irrigating works were planned by eminent engineers; the canal owners have water power and city (Ogden) water works to look after, which will compel them to maintain the works in good repair. The water may be depended upon absolutely. The valley can hardly be excelled in the world for beauty of surroundings, ease of access, convenience to markets, fertility, climate, all that makes land desirable and valuable. Improved, $40 an acre can be realized from it in profits every season by a thrifty and enterprising man. The valley can be made, and some day will be, a garden twenty miles long by ten miles wide. Other irrigating and settlement inducing companies have been organized.

A TYPICAL CASE.—The following statement, based on actual experience, and made by one of the most intelligent farmers in Utah, shows the prime cost of settlement; what he has done and what may be done with a typical farm of forty acres, well irrigated land and properly handled

Expenditure.

First cost 40 acres of land and water right, $40.00 per acre...............	$1,600.00
One mile of fence (4 wire)............	140.00
Dwelling house, complete	600.00
Stable, barn and sheds	260.00
Clearing, plowing and harrowing 40 acres.............	150.00
100 shade trees............	15.00
200 fruit trees	30.00
10 acres planted to alfalfa and seed ...	20.00
20 acres wheat and seed ...	30.00
4 acres of potatoes, seed and planting.............	20.00
5 acres of oats	10.00
Water rental.............	80.00
Total.............	$2,955.00

First year's returns, harvested.

800 bushels of wheat, 60 cents per bushel.......	$480.00
1,200 bushels of potatoes, 50 cents per bushel.............	600.00
250 bushels of oats, $1.00 per bushel	250.00
10 acres of alfalfa and seed (½ return).......	150.00
Total.............	$1,480.00

The above statement shows a net earning of 50 per cent, or one-half the total amount invested, for the first year's work.

VALUE OF IMPROVED FARM LANDS.—Improved farm lands in Utah range in value from $5 to $225 an acre, averaging $40 to $50. They are rapidly appreciating in price, however, and it begins to be seen that every acre subject to water, or that can be made subject to water, is intrinsically worth $100. Lands about Grand Junction, Col., entirely similar in character and not so favorably situated as the lands of Utah and Salt Lake Valleys, five years planted in fine fruit, have been held by their owners against offers of $500 per acre. There is not much public farming land to take up; as a rule it must be bought. It is limited in amount, substantially, removed from the influence of Eastern competition; its market is in the growing towns of this and adjoining Territories, in the mines, and among the manufacturing classes. Products always command fair prices in cash. There are other streams in the Territory where the Bear River Canal irrigating scheme can be profitable duplicated. There is undoubtedly five times the present cultivated area to be reclaimed by the storage of water and turning it upon the thirsty soil in the dry season.

UNION PACIFIC LANDS IN UTAH.—The following is from the Salt Lake *Journal of Commerce* :—

"The work of the Government with respect to irrigation, if properly performed, may lead to the settlement of millions of acres of land in the arid regions of the West which now are lying untilled for the want of a supply of water for irrigation. One only need traverse the rich ranges that rest along the eastern flanks of the Wasatch and note the fertile character of the soil, to be convinced that only water service is needed to make vast tracts, now wholly unoccupied, become most attractive as well as productive. Yet it must not be supposed that all the lands open for settlement lack water supply. In the rich valleys of the northern water-shed of the Uintah Mountains, both in Wyoming and Utah, there remain unfenced and unimproved large districts inviting to the plow and to which an ample water supply can be conveyed easily from unappropriated sources. Other tracts cannot be irrigated, yet make excellent grazing lands, while still others can be converted into fields and pastures if ever the government will reach forth its aid toward the building of reservoirs and canals, which settlers cannot think of contemplating at their own cost.

"Of these lands a large proportion are owned by the Union Pacific Railway Company, and most favorable terms are now being offered to settlers, the terms of payment being easy, and large concessions made to bona-fide purchasers who improve the country. Of such lands, no fewer than half a million acres of the most varied character are for sale in Utah alone; and from this area there are surely some very desirable homesteads to be selected. On the growth and settlement of the surrounding country depends much of Salt Lake City's future greatness, and we would like to see every cultivable acre of this ground under the plow and yielding the fruits of the earth for the sustenance of the thousands who are moving hitherward.

GOVERNMENT AID TO IRRIGATION.—Congress has begun to look into the subject of irrigation in the arid region, and a Washington correspondent of the *Salt Lake Tribune*, reports Captain Dutton, of the U. S. Army, in charge of the field engineering works, as saying:

" There are two ways in which this gigantic problem can be solved: One is that Congress shall appropriate the money to create the system of irrigation required, and that the Government shall then sell the land at its improved value to those who want to buy. In my opinion, the money thus received would more than pay for the expenditure.

" The second practicable scheme is that the Government shall grant to individuals or corporations all the lands that they will redeem, providing that the grantees shall construct works on plans endorsed by the Government, and subsequently dispose of the lands on terms approved by the Government and the department.

" 'And which of these plans is likely to be adopted?' asked your correspondent.

"Neither, so far as I can see. Either is bound to meet with the strongest opposition in Congress — so strong, indeed, that I don't see how it can possibly get through.

" ' Is not much of this water that it is proposed to use the property of private individuals and companies? '

" Yes; and that point is not unlikely to give rise to complications. You see, the United States Government has relinquished control over the waters to be utilized, and, as the law now stands, anybody in the dry region can establish a claim over as much water as he wants by simply filing what is called an appropriation, providing only that he does not interfere with any previous claim. But this trouble, I take it, could be easily gotten over; for it has frequently been decided legally that no one can own running water. Even the canal companies are in law merely common carriers of water, and the rates they charge are subject to regulation.

" 'And the available water, properly applied —? '

" Would do wonders, of course. Vast tracts of land worth to-day 10 cents an acre would be given a market value of from $50 to $100 an acre. The effect of irrigation on the producing powers of any soil is marvelous. In the Mississippi Valley, which has rain and not irrigation, the average yield of wheat is fourteen bushels per acre; the Western irrigated land rewards the husbandman with a harvest of twenty-six to twenty-eight bushels per acre. It is the same way with other crops. Alfalfa, you know, is the great vegetable product of the arid region. It is a plant from the Mediterranean, and the most useful forage known; it will fatten horses, sheep and cattle surprisingly; it is hay, but more than hay — like exaggerated clover, as tall as timothy grass, affording three crops a year, of two-and-a-quarter to two-and-a-half tons to the acre. It is to the West what corn is to the Mississippi Valley, and its future in the agriculture of that part of the country is enormously promising, supposing that irrigation becomes an accomplished fact. The cattle of Uncle Sam's domain are born in the West, and, instead of leaving them to die of starvation in the cold months, as formerly, it has lately become the fashion to winter them on alfalfa, and this has already created an almost unlimited demand for the fodder.

" ' The irrigation proposed is by canals? '

" By canals, presumably, with huge reservoirs in which water will be stored during the spring floods, and from which it will be let out gradually in the dry season."

DAN. DE QUILLE ON THE GREAT BASIN. — "But first let the interior of the continent be reclaimed and settled. The people of the East are now sufficiently strong to build their own irrigation canals and ditches, if they want them. It will probable take some of them fifty or a hundred years to find out that what is good for a garden is also good for a field. It is the great interior that should first receive the attention of the Government— the region in which are supposed to lie the ' lands for the landless.' The lands are there, but the 'landless' will not go to them while they are water-less. Let a proper system of irrigation be established for one large experimental tract, and the landless will be gathered there ready to accept the terms offered settlers, before the cement of the dams is dry. About that reservoir (if the land shall be properly sub-divided) will soon be seen a village settlement as dense as are those below the great tanks of India and Ceylon.

" Millions may (and one day will) find homes in what is now looked upon as little better than a desert. The soil is the best and strongest in America. It cannot be overtasked or worn out. The lands of Montana, Idaho, Utah, Nevada, Arizona, New Mexico and Colorado are capable of supporting a more dense population than any other in the United States, wherever water can be procured.

" With the present and promised railroad facilities, it will be an easy matter to ship all surplus ranch products to either the Atlantic or Pacific seaboard. What is now looked upon as being almost out of the world will soon be seen to be in its exact center.

There is now no other West than that which lies between the Rocky Mountains and the Sierras. The country is now pretty well built up on the two sides. The only opening is in the middle. The people of the Atlantic States who rush to California to acquire property and " grow up with the country " make a great mistake. Many of them found that out when struck in the face with the prices there asked for real estate. If their desire was to grow in wealth with a country, they should have gone to a country not already wealthy. California will do for wealthy invalids who can pay fortunes for an orange grove, in the shade of which to sigh out the other lung; but for men of moderate means, possessed of lungs which they wish to keep, the new and growing States and Territories between the two great mountain ranges is the place.

TYPICAL PLACE FOR A COLONY. — In Southern Nevada a colony of Eastern people, who want a warm climate and immunity from all lung troubles, may buy for a song a region that is even now a little Eden, and which will yet become the paradise of America. It is a land of almost perpetual sunshine, and all know, who know anything of pulmonary complaints, that sunshine and freedom from fogs and damps is what the consumptive or weak-lunged must have to get new life. They must get far away from the sea and swamps and regions of cloudy skies.

In the southern part of Lincoln County, Nevada, on the borders of the Colorado river, where is a grand stream half a mile wide, is a beautiful land large enough for all the consumptives in New England. It is the land of the lemon and the orange —

" Knowest thou the land where the lemon trees do bloom,
And oranges like gold in leafy gloom,
A gentle wind from deep blue heaven blows,
The myrtle thick, and high the laurel grows?
'Tis there! 'tis there,
O my beloved one, I with thee would go!"

It is the land of which Mignon sang to Wilhelm Meister — the land she saw in her dreams.

There side by side grow the olive and the plum, orange and apple, lemon and peach, fig and apricot, pear and pomegranate — a land of the grape, almond and walnut. Also it is a land of cotton, sugar cane and tobacco, corn, wheat, oats and all other kinds of grain. There, too, grow to perfection all kinds of kitchen vegetables—melons, squashes, beans, sweet potatoes, yams, pea-nuts, and everything grown in any place in America, north or south.

Two crops are grown yearly on the same land. It is first sown in small grains—wheat, barley and oats—which are harvested about the first of June; it is then planted in corn, beans, potatoes, beets, cabbages, onions, turnips, squashes, melons and other such garden vegetables.

These lands are on the Rio Virgin and the Muddy. In the valley of the Muddy (it deserves a better name) are many thousands of acres of land which may be irrigated by means of the supply of water afforded by the stream. Over 5,000 acres was irrigated by the Mormons during their occupancy of the country, but by constructing storage reservoirs more than twice that area might be watered. The soil is deep, of a reddish color and very rich. It is filled with lime pebbles, and as the waters of the streams contain carbonic acid these pebbles are slowly dissolved and feed the soil. All the country is limestone. In 1869 there were fifty Mormon families on the Muddy. At Los Vegas Valley there were some twenty-five families, and others in other valleys in the same neighborhood. At Los Vegas bursts up out of the limestone a spring that pours out the year round a stream of 350 miner's inches of water. Near by, on the Rio Virgin, is a great mountain of rock salt, so transparent that a newspaper may be read through a block of it a foot in thickness.

In the fall of 1870 all the settlers were ordered out of this region— were made to return to Utah by command of Brigham Young. Nearly every family obeyed, but it was with weeping and wailing that they left their beautiful homes in this paradise of the Great Basin. After their departure all went to ruin. Houses and fences were burned and destroyed, and all grew to be a sort of semi-tropical wilderness.

There are now only a few settlers in all this beautiful region. A colony could buy their right for a small sum. They are too few and too poor to help themselves. Of all that the soil can produce they have a superabundance, but they are hundreds of miles from any market. A railroad would place them in the center of the world, almost. They could dash down to Los Angeles, "by the sea," in a few hours, or in a few hours could be in Salt Lake City.

A colony of New Englanders in this garden spot of the Continent would have means and influence to at once have the Salt Lake and Los Angeles railroad put through. While the road was building they could have a little iron propeller running up and down the Colorado, for that river is navigable and free of all obstructions (with a great depth of water) all the way from Callville to the Gulf of California.

From Callville the settlers would have only a short distance to transport their goods by team, and all the way over a level road.

The mouth of the Rio Virgin on the Colorado is only 800 feet above the level of the sea; the valley of the Muddy, 1,115 feet; Los Vegas, 2,095; Hiko, 3,700. Almost any desired elevation is to be found.

On the whole continent no better place is to be found in which to establish such a colony as has been mentioned. All, too, would speedily grow rich, as well as sound in health. It would be a very different thing from planting a colony in a desert, as has been tried in Lower California. My colony would be composed of well-to-do folk — people able to plant themselves.

To reach their destination the people would go to Needles by rail and then take their own little iron steamer up the Colorado to Callville. However, let them go to their little domain by team or on foot, the indications at present are that there would be a railroad passing by as soon as they had got comfortably settled and had anything to send to market. No other region would be so near to the north with tropical fruits by several hundred miles.

There are a thousand industries in this great interior region to engage the attention of the industrious settler — scores of which are unknown in the East. In the older States a mountain is a pile of rock and earth out of which nothing of value can be obtained, and a valley is a patch of corn or potato ground, and nothing more. Here nearly every mountain contains the precious or the useful metals, and in the valleys are bottomless beds of minerals. At the same time, the soil of both valleys and mountain slopes is the strongest and most productive in the world, being composed of lava and other volcanic rocks that decomposing form a soil not only virgin but also rich in all kinds of mineral conducive to the growth of every kind of vegetation. It is well known that on the slopes of Vesuvius, in a soil consisting almost wholly of decomposed lava, are some of the finest vineyards and most beautiful and productive gardens in the world. Our Great Basin region is all Vesuvius slope. All it requires is a little water on the "slope." Put on the water and at once is seen an "eruption" of vegetation.

STOCK AND SHEEP.

Statement showing number of horses, mules, asses, cattle and sheep assessed in Utah Territory for the years 1890, 1891 and 1892, and the assessed value for 1892.

HORSES AND MULES.

Counties.	1890.	1891.	1892.	Assessed Value.
Beaver	2,574	2,513	2,634	$ 76,182
Box Elder	4,477	6,724	*	112,225
Cache	6,262	7,869	7,980	321,950
Davis	5,060	3,145	8,100	132,145
Emery	2,626	2,616	2,962	106,750
Garfield	2,032	1,208	3,438	89,745
Grand	1,131	1,504	1,488	36,030
Iron	1,875	2,202	2,082	71,790
Juab	2,097	2,093	2,081	71,285
Kane	2,145	3,160	2,791	85,143
Millard	4,891	3,662	2,792	72,980
Morgan	1,233	1,329	1,383	44,225
Piute	2,464	2,053	1,310	31,724
Rich	2,366	2,174	2,380	83,264
Salt Lake	....	8,438	7,060	309,175
San Juan	791	1,035	1,044	30,000
San Pete	5,002	4,995	5,832	200,610
Sevier	3,902	2,790	4,126	126,940
Summit	3,308	2,911	2,982	126,587
Tooele	3,682	3,704	5,016	103,728
Utah	6,781	7,857	8,403	329,205
Uintah	3,149	3,542	3,836	92,079
Wasatch	2,000	2,360	2,387	86,085
Washington	1,965	2,397	2,635	88,915
†Wayne	....	....	3,119	42,671
Weber	4,082	4,498	4,646	213,040
Total	75,895	85,579	87,457	$3,084,473

*No report. †New county; no report before 1892.

CATTLE.

Counties.	1890.	1891.	1892.	Assessed Value.
Beaver	6,392	6,740	9,287	$ 99,217
Box Elder	10,094	13,297	*	147,498
Cache	9,968	12,913	11,937	142,510
Davis	9,538	5,530	5,254	78,594
Emery	9,707	5,530	15,001	112,305
Garfield	9,024	6,256	15,279	140,128
Grand	23,543	19,593	17,513	175,490
Iron	6,706	8,968	8,315	98,142
Juab	2,790	3,117	4,734	46,390
Kane	9,801	12,949	13,655	179,162
Millard	6,206	8,605	5,988	60,085
Morgan	3,547	3,333	3,635	39,170
Piute	9,415	4,582	2,063	20,630
Rich	9,307	8,503	8,489	91,006
Salt Lake	....	8,128	6,587	136,807
San Juan	27,392	26,362	29,722	294,720
San Pete	9,711	10,161	10,958	125,485
Sevier	10,513	14,719	10,067	104,908
Summit	8,845	8,966	8,364	113,879
Tooele	4,844	5,061	6,071	80,145
Utah	12,013	12,059	13,883	177,150
Uintah	11,494	9,469	8,791	70,941
Wasatch	9,383	10,211	6,917	76,500
Washington	10,402	10,209	17,329	195,980
†Wayne	....	....	8,582	85,820
Weber	6,841	6,976	7,274	103,720
Total	237,496	242,235	255,675	$3,000,872

*No report. †New county; no report before 1892.

SHEEP.

COUNTIES.	1890.	1891.	1892.	Assessed Value.
Beaver	48,061	97,826	35,567	$ 71,134
Box Elder	80,215	97,598	*	90,285
Cache	4,010	1,758	8,521	17,468
Davis	4,962	10,783	4,703	7,916
Emery	156,440	21,410	75,695	166,670
Garfield	16,311	26,402	26,775	53,550
Grand	....	14,000	9	18
Juab	132,220	148,611	57,257	114,451
Iron	41,642	48,967	54,080	106,980
Kane	85,346	96,025	49,740	99,480
Millard	180,088	190,000	48,075	96,150
Morgan	947	4,568	4,407	6,506
Piute	27,440	36,785	11,944	23,885
Rich	4,201	6,640	*	....
Salt Lake	....	201,536	5,788	8,279
San Juan	8,100	9,850	12,200	18,305
San Pete	2,423	105,136	216,272	432,544
Sevier	31,967	22,989	58,258	112,518
Summit	8,304	4,923	6,285	12,035
Tooele	189,088	162,469	187,167	365,481
Utah	63,347	101,605	74,306	148,280
Uintah	41,115	41,165	37,425	56,168
Wasatch	9,322	10,731	15,000	30,000
Washington	11,843	14,680	15,212	30,425
†Wayne	....	....	21,495	42,990
Weber	4,903	18,990	20,799	41,616
Total	1,150,295	1,485,392	1,045,080	$2,153,107

* No report.
† New county; no report before 1892.

Increase for the year:

	Number.	Per Cent.
Horses and Mules	1,876	.2
Cattle	13,440	.5
Sheep	440,312	29.6

Wool clip for the year (estimated), lbs..12,000,000
Number of cattle exported (estimated)..42,000
Number of sheep exported (estimated)..650,000

The corporation formed for the purpose of establishing a stock yard near Salt Lake City have erected the necessary buildings, and are now receiving and shipping stock.

Inclusive of desert, bench land, mesa and mountain side, there are, perhaps, 10,000 square miles of cattle and sheep lands in Utah, sustaining 300,000 head of horned cattle, 100,000 horses and mules, 1,500,000 sheep, and other animals in due proportion. (The assessors' rolls do not show quite so much, but that is easily accounted for; the real figures would show even more than the foregoing.)

Many Utah men graze their flocks and herds, besides, in adjoining Territories, and quite a percentage of the wool shipped away by Utah is grown by men who live over the border, but who market their clip from Utah. The wool crop of 1891 was over 12,000,000 pounds. The average per fleece is about six pounds. The output of beef cattle and mutton sheep yearly is very considerable.

GOLD AND SILVER MINING.

OUTPUT TO DATE.—From the commencement of mining in Utah, in 1871, to the close of 1891, twenty years, the total output of silver, gold, lead and copper, rating silver at its coining value, as the U. S. Mint officers always do, and lead and copper at their average yearly price in New York, reaches, in round numbers, $175,000,000 in value.

TABULATED STATEMENT.

	Refined Lead.		Unrefined Lead.	
	Amount. Pounds.	Value.	Amount. Pounds.	Value.
1879	2,301,276	$ 103,557.42	28,315.859	$ 592,095.57
1880	2,892,498	144,624.90	25,657,643	641,444.75
1881	2,645,373	145,495.51	38,222,185	955,554.62
1882	8,213,798	410,690.00	52,349,850	1,361,096.00
1883	3,230,547	161,527.00	63,431,964	1,585,799.00
1884	4,840,987	169,434.54	56,023,893	980,418.12
1885			54,318,776	1,222,176.46
1886	208,800	9,667.44	48,456,260	1,405,231.54
1887	2,500,000	111,750.00	45,678,961	1,196,788.77
1888			44,567,157	1,203,313.28
1889	2,359,540	89,662.52	59,421,730	1,378,584.13
1890	5,082,800	203,312.00	63,181,817	1,895,454.51
1891	6,170,000	246,800.00	80,356,528	2,410,695.84
Total	40,445,619	$1,796,521.23	657,982,123	$16,828,652.54

SILVER, GOLD AND COPPER.

	Silver.		Gold.		Copper.	
	Amount. Ounces.	Value.	Amount. Ounces.	Value.	Amount. Pounds.	Value.
1879	3,732,247	$4,106,351.70	15,732	$298,908.00		
1880	3,663,133	4,029,501.30	8,020	160,400.00		
1881	4,958,345	5,503,762.95	6,982	139,640.00		
1882	5,425,444	6,114,874.00	9,039	180,780.00	605,880	$ 75,735.00
1883	4,531,763	4,984,939.00	6,991	139,820.00		
1884	5,569,488	6,123,017.04	5,530	110,600.00	63,372	6,337.20
1885	5,972,689	6,221,596.56	8,903	178,060.00		
1886	5,918,842	5,860,837.34	10,577	211,540.00	2,407,550	144,453.00
1887	6,161,737	5,976,884.89	11,387	227,740.00	2,491,320	124,566.00
1888	6,178,855	5,787,527.51	13,886	277,720.00	2,886,616	288,681.60
1889	7,147,651	6,656,254.65	24,975	499,500.00	2,060,792	206,079.20
1890	8,165,586	8,492,209.44	33,851	677,020.00	956,708	76,536.64
1891	8,915,223	8,759,206.59	36,160	723,200.00	1,836,060	100,983.30
Total	76,451,053	$78,616,993.97	192,083	$3,824,928.00	13,308,498	$1,023,371.94

Increase over 1890— | | | | | | Per Cent.
In pounds of unrefined lead... 21.18
In pounds of refined lead... 21.38
In ounces of silver 9.18
In ounces of gold... 6.82
In pounds of copper.. 90.87

DIVIDENDS OF 1891.—Mines earned dividends in 1891 as follows:

Bullion Beck..$330,000
Mammoth 280,000

The first two mines are close corporations which do not declare dividends. Their earnings are set down upon best information obtainable. Utah mines have paid in dividends to date about $25,000,000.

WHERE THE MINES ARE.—The mines wrought at the present time are mainly in Beaver, Juab, Summit, Salt Lake, Tooele and Washington Counties. The northern mines lie on the same parallel in Tooele, Salt Lake and Summit Counties. The mines of Juab County are eighty or ninety miles south of these. Beaver County is 200 miles and Washington County 300 miles south of Salt Lake City. Mines were wrought to some extent in Wasatch, Weber, Box Elder and Piute Counties. There is, in fact, no county in the Territory where the prospector has not left his footprints.

Wherever, in Utah, there are mountains, mineral indications are not wanting, and valuable minerals are likely to be found in time in paying veins or deposits. Ores of good quality are known to exist in many of the isolated ridges which break the face of the desert in Western Utah, but mining in that section still awaits the construction of railroads.

The main producing district of the Wasatch Range lies on the heads of the Cottonwoods and of the American Fork, within sight of Salt Lake City, and over the ridge eastward, where the waters find their way into the Weber and Provo Rivers.

Northward from this locality nothing of importance has yet been found, but two hundred miles southward, on the head of the Sevier River, eastward of the town of Beaver, there is a district called Marysvale, containing some promising mines, the development of which is retarded by the comparative isolation of the district.

Mines are found on both flanks of the Oquirrh Range, from Great Salt Lake southward nearly one hundred miles, as at Stockston, Dry Cañon, Ophir, Bingham and Tintic. All these localities, except Marysvale, are connected with Salt Lake City by rail. The mines of Beaver County are at Frisco and about Milford, the terminus at present of the Union Pacific Railway. The mines of Washington County occur in a sandstone reef which extends along and near the base of the Wasatch for one hundred miles.

BEAVER COUNTY.—Beaver County contains four or five parallel ranges or ridges, striking north and south, and all of them mineral-bearing. A single chimney of ore in a contact along the east base of Grampian Mountain (Horn Silver Mine) turned out 90 tons of ore a day for four years, realizing to its owners more than $13,000,000, $4,000,000 of which was disbursed in dividends. After this enormous output the mine had three or four hard years, but is again doing well. Ore bodies have been opened in new ground on different levels. Shipments for the year 1891 were 21,160 tons.

A few years ago two French companies were at work on some promising copper mines about three miles northwest of Frisco. They ran some copper bullion; but later on, for some reason, probably lack of sufficient means in hand to properly open the mines and put in a smelting plant, they shut down and quit work. There is not much doing about Frisco aside from the operation of the Horn Silver Company.

In Star District, south of Milford, two or three men, more persevering or more fortunate than the mass, have gone through the pinch or fault that cuts off all the surface deposits of the district within 150 feet of the surface, found their veins again, as full and rich as ever, are shipping ores and prospering. Notably is this true of the Talisman and Stalwart, the developments in which not only prove it to be a valuable mine, but have infused new hopes into the owners of other mines. So that there is good prospect of the revival of the activity which prevailed in Beaver County mines in the seventies. Amongst the mines upon which work has been resumed and which are increasing their output, are the Atlas (old Wascoe), the Rose, formerly the Mazeppa, the Anvil, the Creedmore, the Monitor, the Burning Moscow, the Rebel, the Maringo, the Lady of the Lake, the Mammoth, the Elephant, and others. The ores of all these mines are rich above the average of Utah ores. There are a number of other districts within twenty miles of Milford which the resurrection of Frisco and Star will doubtless recall to life.

The facilities for mining in Beaver County are very good. The country is dry in the summer, but there is sufficient water, wood, and timber for mining purposes, and operations are not obstructed by snow or cold in the winter. The ores are carried from the mine dumps by wagon and rail to the Salt Lake smelters at about $7 per ton. The mines are easy of access. Provisions and supplies are cheap and abundant, and good labor is obtainable at fair rates of compensation. There can be no doubt that extensive exploitation and operation in this county would very greatly increase its output. "Old Baldy," a huge mountain overlooking Beaver Valley on the northwest, contains numerous gold fissures, some of which have been mined with more or less profit; and gold has recently been found in paying quantities at Indian Creek. At Cove Creek is one of the largest and best sulphur mines in the world. It is constantly producing and shipping.

JUAB COUNTY, TINTIC.—Tintic is the principal mining district of Juab County. It is in or on the western slope of the Oquirrh Range, which here rises perhaps 2,000 feet above the general level of the country making the absolute altitude 6,000-7,000 feet. A correspondent of a Salt Lake paper gives the following excellent description of the country:

"The mountain mass in the Tintic District is composed almost wholly of folded up strata of limestone, which is also the mineral producing formation of the region. A thick bed of quartzite extends along the slope of the Range, but the higher central portion of this has been removed by erosion, leaving a broad belt of the underlying limestone exposed.

. "North of Eureka Gulch some eruptive rock (probably trachyte) is also found, covering a limited area, and one prominent peak is composed of this material. But this formation is not very extensive, and I believe is the remnant of a once intrusive sheet (lacolite) between the limestone and quartzite.

"Crossing the Range then from west to east along the transverse ridge of which Eureka Hill forms the central and highest point, and which furnishes an excellent cross-section, we pass over a continuous limestone formation for a distance of a mile or more of strata standing nearly vertical.

"This distance represents a number of folds, or, rather, I think three anticlinals and two synclinal folds, for I doubt if the limestone strata is more than 1,000 feet thick, although during my short stay I could not enter into details of accurate measurement.

"The ore deposits are found in vertical sheets or veins, very irregular, faulted and 'pockety,' as limestone deposits always are, yet following the strata of the rock, and consequently have a strike of nearly north and south, and probably a very slight general dip toward the east.

"These limestones are highly crystaline and very hard, ranging in color from a grayish white to a dark blue, the latter greatly predominating. For lack of lithological data and proper time to investigate, I am not prepared to give their geological position, but should judge them to be very old, and probably belonging to the lower carboniferous or even older.

"I had been told that the ore deposits of Tintic are identical with those of Leadville, Colo., with which I am quite familiar, and this is to some extent true; but they differ in some essentials, the most important of which, to the practical miner and prospector, is their mode of occurrence.

"The deposits, as before stated, being vertical, tunneling would be in order, but the country being quite flat, does not permit of this to good advantage; and the most common way of prospecting therefor is by shaft and cross-cuts. But as there is no outcrop, it is not easy to determine where to sink in order to obtain good results.

"Two theories may be advanced regarding the manner in which these ore deposits were made, either of which, or both combined, would probably have given the existing results:

"First—The ore may have been deposited in and along the top of the limestone floor before the disturbance, and carried along, taking its proper place in the folded-up strata, probably being altered to some extent later.

"Secondly—The deposits may have been made altogether after the disturbance, as a fault-fissure, by far the most common form of vein. In the first case we may expect the ore to continue to the bottom o. the limestone and there break off abruptly. In the second it may continue downward far beyond the limestone and into the underlying rocks.

"In either case, I have little doubt that the ore deposits follow along or close by the contact plane of the synclinal fold, and this would tend to prove the first theory, but it does not disprove the second, as the folding plane would most likely also be the plane of faulting and rupture and consequent mineral secretion.

"It is not easy to distinguish on the surface the plane of contact between the various folds, but to the practical and careful observer this can be done and may greatly assist prospecting.

"During the long period this enormous limestone bed was being deposited on the bottom of the sea, the conditions varied slightly; hence the composition of material also varied from the bottom to the top, and especially so in the amount of silicious matter it contains. Some being highly silicious, of a light color, and very hard, while the bulk is more soft, darker in color, and contains less silica, and these conditions still exist.

"A thin layer of shale was also formed toward the top, which can be seen in various places and may be used as a landmark.

"The ore deposits are as a rule large, easily mined, and of a high grade. The Bullion-Beck, Eureka-Hill, Centennial-Eureka, Crismon-Mammoth, and a few others are the principal mines, all heavy producers and dividend-payers, but aside from these little prospecting has been done. A small number of claims have been patented; a few more are held by location. For miles in each direction the country is practically virgin ground.

"Owing to the low altitude the winter snows depart early, leaving the ground parched and dry. Vegetation is very scant, and timber for mining purposes has to be brought from other parts.

"The mines are not troubled with water, but rather with the lack of it, for even those mines which have reached a depth of nearly a thousand feet have none excepting what is brought there by human effort and ingenuity. Most of the water for all kinds of uses is derived from springs, which seem to be quite numerous in certain places, but their flow is not strong, and they are already taxed to nearly their full capacity.

"As the population of Eureka and the various other camps is steadily increasing and more mines are being opened every year, the question of water supply cannot be far distant, for when the section receives the attention it surely merits the few local springs will be far inadequate. But Utah Lake being distant only a few miles may be counted upon to supply the means; the end can be easily found."

Another observer, namely, Mr. J. E. Rockwell, of Pueblo, Col., has the following to say of the Tintic District:

"I spent five days there. I found a lime belt, well defined, standing on edge, from half a mile to three miles in width, and many miles in extent. The line of porphyry on the east and the quartzite on the west is clearly marked, and between these two great walls there is a vast lime zone. It is best illustrated by a book standing on its back with its pages turned up. Between these leaves extend, in a course generally north 15 degrees east, great channels of ore. In sinking down between any of these layers of lime, one is liable to find ore, and when found the bodies are extensive—sometimes breaking off from one layer of lime into the adjacent, but always at some place connected so as to be easily followed. Necessarily this ore, through leaching, would be found at some depth from the surface, and for this reason it is not a poor man's camp. But it possesses the advantage of great certainty of finding ores to one who is able to sink to the requisite depth. The Eureka is sunk to 1,000 feet and the Bullion-Beck over 600. The latter company are now erecting a $300,000 plant and preparing to sink to a considerable depth below their present workings. From these shafts levels and cross-cuts are run in all directions at different levels until ore bodies are discovered. It has been demonstrated that the quality of ore is high grade and continues in depth. It is typically a high grade camp, although quantities of low grade ore are necessarily found, which some day will be worked.

"The mountains surrounding the place are easy of access, and there is every facility for mining operations, which can only be carried on by companies and men of means.

"It is assured that vast wealth is stored in this lime belt only awaiting development."

There appears to be three main ore channels, one to three miles long and half a mile apart, and these are located under various names. Output is all that counts in mining, and the leaders can be detected by amount of ore shipped, given in table below.

The Eureka Hill was for years the only mine of real note in the district, but although it doubled its customary output last year, it was far outstripped by the Bullion-Beck. During the past five years the Eureka Hill has extracted and sold 85,000 to 90,000 tons of (probably) $50 ore. If mining cost $10; freight to market, mainly in Colorado, $8, and smelting $12 per ton, the dividends of this five years would be very large, as can be seen. The company is a close corporation and dividends are not published. In December, 1888, some proceedings brought the company into court, and the books showed the payment to that time of 124 dividends of $10,000 each, equal to $1,240,000. This is about $250,000 a year for the five years of full and steady output. The past two years have seen the output increased, but more has gone into improvement. About $150,000 was expended on plant in 1890, and the property has now an excellent equipment. The main shaft has reached the 1100-foot level exploitation is kept well ahead, and there are always large ore bodies in reserve.

The Bullion-Beck, like all the successful mines of the district, struggled along empty-handed for years, until within two or three years. In 1889, expensive suits with the Eureka Hill, which it adjoins, were compromised, and its net earnings are said to have been $375,000. In 1890, besides expending $90,000 on a hoist, purchasing Homansville springs and the Deprezin group of eight full claims at $50,000, it paid seven dividends aggregating $325,000. In 1891, it paid $330,000. Before the purchase of the Deprezin group the company owned about fifty acres. This purchase makes their holding upwards of 200 acres. The working shaft has reached the 700-foot level, and ore is being extracted from the second to the sixth levels inclusive. The mine has a great future.

The Centennial-Eureka lies south of the Eureka Hill, on the great ore channel between the Eureka Hill and the Mammoth. It comprises ten or twelve claims, and extends along the ore channel 3,800 feet. The owners have had difficulty enough to get an insufficient hoist up on the hill, to sink a shaft 500 feet, and to extend the upper levels southward. But they have their reward at last. Down to January 1st, 1890, they had sold but 1,365 tons of ore, the total output of five years. Last year they put out 3,667 tons, and sold 3,396 tons for $508,669, almost exactly $150 a ton.

They paid upwards of $100,000 for mining ground, paid $150,000 in dividends, and had $172,000 left on hand. Last fall some time, the 300-foot level struck a chimney or ore-shoot far south of the shaft, which might be likened to Aladdin's cave for richness, and which proved to be 200 feet long on the strike of the vein. None of this ore has been marketed as yet. The mine needs a new working shaft and hoist, but it has the wherewithal to get it.

The Crismon-Mammoth lies over on the other (southeast) side of the high point up which the Centennial-Eureka crawls, and a mile or so distant. This is a great mine, too, and has had vicissitudes. In the course of about twenty years it had paid thirteen dividends, amounting to $210,000. About two years ago the 600-foot level was driven north from the working shaft about 1,000 feet, along a contact between two different lines. It developed an ore-shoot or chimney about 100 feet through on the strike of the vein, and from 4 feet to 16 or 20 feet thick, from which 9,500 tons were shipped during the first year. The superintendent, Captain H. H. Day, who died toward the close of the year, called it an average of 50-ounce (silver) ore, carrying 13 to 15 per cent lead. But out of the 9,500 tons shipped $560,000 in dividends were paid, and this would seem to indicate a much richer ore, something like $75 ore.

All these mines have to pipe and pump water from a distance to make steam, etc.

The Gemini group includes, with several other claims, the Excelsior, Keystone, and Red Bird; the territory is 5,400 feet in length. The owners are sinking shafts on the three claims mentioned, and opening levels from one to the other. Very good ore is coming out of the Keystone from the third to the sixth levels inclusive, and there is a great deal of ground ready for stoping.

The Eagle group comprises nine claims east of and contiguous to the Eureka Hill, which, during 1890, shipped 500 tons of ore.

The Northern Spy, after an output of $400,000 above the first level, in May, 1890, was sold for $80,000, to persons connected with the Bullion-Beck. The mine produces both milling and smelting ores. A first-class 10-stamp chlorodizing mill, located at Homansville, belongs to the company. On the first level the vein pinches, but the ore is coming in again between the third and fourth levels. The ore is high grade, even for Tintic.

The Godiva group has been worked, off and on, for a dozen years. It comprises several claims, springs, and a well-site, and is situated a mile or two east of the Eureka Hill, at the extreme northeast end of Mammoth Mountain, above Burriston Pass. The vein is large and strong; the country is limestone; the fact that the ore carries gold, $25.00 and upwards to the ton, and no silver, demonstrates that there is a gold belt in the district.

The Yorkville group, comprising eighty acres, is a mile north of the Eureka Hill, and developments made indicate that the big ore channel is fertile.

There is a great revival in Tintic. New men are constantly going in, the old claims which laid fallow for twelve or fifteen years have been examined, many of them leased and bonded and sold to men who have opened them. Many new locations have been made, and good judges believe that this is but the beginning — that, in short, the mineral belt of Tintic, the contact between the lime and the quartzite, is as yet mainly virgin ground. Amongst the mines and groups upon which good work is now being done, are the Madera Consolidated, the Marion Consolidated, the Plutus, the Sioux group, the Snowflake, the Governor, the Iron Blossom, the Wolf, the Cave, the Hungarian, the British, Copperopolis, the Undine, the Sunbeam, the Treasure, the Tesora, the Turk, the Eastern and Daisy, the Hard Winter, the Belcher Consolidated, a group of eight claims, the Lucky Boy, the Alamo, the Golden Ray, a group of six claims, the Isona, the Retribution, and many more "too numerous to mention."

Many of these Tintic mines, and not the least the mines about Diamond, have immense outcrops, nearly covering the full surface area. The ore is found in bunches and chimneys; greater depth will surely show concentration in large bodies. Experience has demonstrated that these mines as a rule need only to be opened and wrought to become profitable.

For 1891, Bullion-Beck shipped 6,945,680 pounds, which sold for $340,057.04.

Mammoth (1891) shipped 7,650 tons.

Eureka Hill (1891) shipped 19,400 tons.

DEL MONTE DISTRICT.—This district is four miles north of Eureka.

The railroad passes within four miles of the mines which are immense bodies of lead ore, carrying about three ounces of silver per ton and a large percentage of iron. The more these mines are exploited the larger and cleaner appears to be the ore; more than 1,200 feet of openings have been made in ore, which is from 10 to 45 feet in thickness.

WEST TINTIC.—This district is in Tooele County, but one goes there from Eureka, the capital town of the Tintic mines, and so an account of it is given here condensed and compiled from a letter written by Mr. Frank Burk, and published in the Salt Lake Tribune.

Mr. O. P. Rockwell drove Mr. Burk from Eureka to his ranch on Cherry Creek, near the great desert, a distance of 28 miles, crossing the divide between East and west Tintic; and the next morning four or five miles from Rockwell's Ranch to the mines. The mineral belt is very wide, the country low smooth hills. The first mine seen is the Silver Star, owned by John Fleming, upon which a 26-foot shaft has been sunk, and ore taken out which is 47 per cent lead and carries 13 ounces silver and $7 gold per ton.

The Scotia belongs to Messrs. Van Horn, Baskin & James, of Salt Lake. In early times it put out 2,400 tons of ore, which bore hauling to the Salt Lake smelters. Mr. Burk went down about 60 feet and found free gold and ore rich in silver cropping out of the walls, roof and floor. With railroad facilities this would be a valuable mine.

The old Alabama, now called the Midgley, is a patented claim owned by the Mechanics' Mining & Smelting Company of Omaha. A 60-foot incline shows iron carbonate which assays high in silver and gold. Ore was formerly wagoned from this mine to the Salt Lake smelters, but ceased on account of the depression in price of lead and silver.

A mile and a half north of the Scotia is the Northwestern, owned by O. P. Rockwell. An incline exposes a large body of soft carbonates similar to the early Leadville find. Assays give 30 ounces silver and 44 per cent lead.

The Little Chief, owned also by Mr. Rockwell, is situated about one mile east of the Northwestern, and an incline of a few feet discloses a large body of iron carbonates, carrying silver, lead and gold. On the same property a shaft has been sunk 130 feet, which in the course of development produced ore carrying silver, gold, lead and copper. A mill run of this ore yielded 25 ounces silver, $3.60 gold, and 47 per cent lead.

The deepest working in the district was on the Stonewall Jackson, and was 250 feet. This claim is owned by Biddlecome, Hunter, Bartlett and Lee. There is a substantial whim and other machinery and conveniences for hoisting ore. There are several hundred tons of ore piled on the dump. Ore of good quality has been followed from the grass roots, and if railroad facilities could be had, this would be a very valuable mine. Assays show 640 ounces silver and $11.40 gold per ton.

The 88 is situated about one and one-half miles south of the Scotia. It is owned by O. P. Rockwell, one-half, and Marcus Howard and Warren Lewis the other half. A shaft 75 feet in depth discloses a very fine galena ore on every side as well as in the bottom. Forty-five feet from the surface a drift has been run north and south, which shows galena-silver ore on every hand. Assays from this property show 18 ounces silver, 46 per cent lead, and a small amount in gold.

Three-fourths of a mile from and northeast of the 88 is situated the Virginia, owned by Rudolph Hunter of American Fork. Drifting from bottom of shaft about thirty feet below the surface is being prosecuted. The character of the ore is a lead carbonate running high in silver.

The shaft on the Brunswick, close to the Stonewall Jackson, was only down a few feet, but assays running away up in the thousands have been had from the vein which has been followed from the surface.

South of the Brunswick, and running in a southwesterly direction, is a low ridge, on which a vein of white quartz carries free gold, while another of lime carries steel galena rich in silver, and still another is almost a pure or native copper, carrying silver and gold.

An old abandoned property north of the 88, and known as the Grand Cross, has a shaft all in ore. Specimens taken from the dump assayed 165 to 325 ounces of silver. The property is owned at present by Charles Crane, Smith and Howard, one half, and Biddlecome & Co. the other half.

The Tribune, on the same ridge, owned by Charles Crane, Mark Howard and others, had a shaft of thirty feet and showed a vein of something like ten inches, carrying very high grade ore, the character of which is a copper carbonate and steel galena, carrying silver and gold.

THE DESERT.—Fifteen miles west of Rockwell's ranch is a bold and rugged granite mountain, rising from the desert, as an island rises from the sea. This is known as the Desert mining district. On the west end of this mountain are some copper mines of great value. The Copper Star and its extension, the Red Bird, have been bonded recently for $50,000, by the owner, O. P. Rockwell, to Joe Biddlecome and partners. The latter are now working the property and are shipping by way of the Mammoth mill at Tintic. This is a true fissure with solid granite walls, the shaft being all in ore.

About fourteen miles north of Desert Mountain is a long, dark ridge running north and south, and known as Death Cañon mountain. On the southwest end of this mountain, on a spur which runs out from Death Cañon, is situated one of the largest and richest silver-lead mines in the whole West. It is known as the Mammoth mine, and is owned by James Chipman, of American Fork, and O. P. Rockwell, of West Tintic. A tunnel is all in ore which is a lead carbonate, assaying 46 per cent lead and 17 ounces silver. A winze was sunk from close to the mouth of the tunnel, which disclosed a 23-foot vein of ore. The Oasis, Columbia and Drumm mining districts are in the same neighborhood with those already sketched, and the proposed Deep Creek Railroad should consider this route for the railroad to Clifton and Deep Creek. There is ore enough here in sight now to tax the carrying capacity of any single track road in the United States. The road could be run by Ophir and down the west side of the Oquirrh Range, around the south side of Boulder, by Eureka, down East Tintic Valley, and, turning westward, a straight or air-line would carry it through the mines of West Tintic and the districts mentioned, by North Dugway, Clifton and on to its destination at Fish Springs and Deep Creek, where some enormously rich "finds" have been made the past year and in great number.

By this route ore will be found on nearly every mile, while water is found in sufficient quantities and at close enough intervals to answer the purposes of any railroad.

SUMMIT COUNTY.—The mining field which begins on the heads of the Cottonwoods and of American Fork, within sight of Salt Lake City, and extends ten miles over the first ridge of the Wasatch, eastward, is thrown by the winding mountain crests which culminate in that vicinity into four counties. The more important, however, are known as Uintah Mining District in Summit County, and as Blue Ledge Mining District in Wasatch County. These are in reality one district, divided by a geographical county line to which the mineral veins pay no attention.

PARK CITY, a town of 6,000 inhabitants, connected with Salt Lake City by the Union Pacific Railway, is the mining town of the district. It is an incorporated town and has a city government, no indebtedness, and plenty of money in the treasury. The city has a newspaper, a brick city hall and firemen's headquarters, jail, etc., three hose carts and a hook-and-ladder truck, three volunteer fire companies and a large number of hydrants. The water-works system, operated by gravity pressure, has proven ample in all cases of fire since the system was put in, but it is now proposed to bring in a larger supply and better water from Highland Lake, forty feet deep and ten acres in area, lying 2,000 feet higher than the city and five miles distant. There is a light, heat and power company with a 900-light plant, and this is about to be doubled in capacity. There are four churches, three or four schools, two Odd Fellows' and one Rebecca lodges, one lodge each of the Masons, the A. O. U. W. and the Knights of Pythias; a bank with a capital of $50,000, and a very well supported opera house. The absolute altitude of Main street (at the hotel), which has a grade of 300 feet to the mile, is about 7,500 feet above the sea. The streets have been graded and otherwise improved till they are very good. Three or four gulches join each other at the head of Main Street and a little above, and up these, rising in two miles 2,000 feet, are the mines. The Mackintosh sampling mill is at the lower end of this street, near the depots. The Crescent concentrating and sampling mill and smelter, and the Marsac (Daly) thirty-stamp chloridizing mill, are in the town, while the Ontario forty-stamp chloridizing mill stands at the head of the main street. The Ontario mine is a mile and a half up Ontario Gulch south of the mill.

THE MINES.—The Ontario vein for 4,500 feet on its course is owned by the Ontario Silver Mining Company; for 1,500 feet next westward by the Daly Mining Company. The next 2,800 feet, going westward, is owned by men interested in these two companies. Here the Anchor Mining Company takes the vein for 12,538 feet. The latter company put a shaft down 600 feet, near the east end of their property, which crosscuts a fine vein of ore, supposed to be the same as the Ontario, or a parallel vein of similar strength and quality, just below the fifth level. Drifts from the fourth and sixth levels also disclosed the vein, where, from its general dip, it should be. From the Ontario westward, the ground gains in altitude, so that the sixth Ontario level is the Daly eighth, and the Anchor seventeenth. The Anchor Company have a drain tunnel 6,600 feet long, intersecting· the shaft on its twelfth level. The Ontario Company are now driving a drain tunnel about three miles long to intersect shaft No. 2 on its fifteenth level. Extended along the vein to the Anchor it would be nearly five miles long, and take the Anchor water to the twenty-seventh level. If the Anchor has the Ontario or an equivalent parallel vein, then the vein is about 9,000 feet long; and if the vein extends through the Anchor ground it is 20,000 feet long. There is good reason to suppose that it continues westward to the Cottonwood mines, and that it strikes eastward through Blue Ledge District via McHenry Gulch for about two miles, making in all six or seven miles, throughout which, with intervals of barren ground of course, it may reasonably be expected to be fertile. It is now claimed that the district has four or five parallel veins.

The Ontario mine is the leader in the extent·of its operations, in cost of plant, in output and dividends. There are upwards of 30 miles of openings in the mine, and about 160,000 cubic yards have been stoped out to get the (in round numbers) $27,000,000 which the mine has produced. The mill and mine plant cost $2,700,000, and mine and mill give direct employment to between 400 and 500 men at an average wage of $100 per month, and indirect employment to a great many more. During the past year the output was 24,694 tons of ore. The gross sum received for the product of this ore was $1,892,421.77; out of which $900,000 was paid in dividends.

Disbursements of 1890 were as follows, those for 1891-92 not varying greatly:

Pay roll and salaries	$535,000.00
Cord wood	34,180.75
Lumber and timber	35,649.50
Coal (from Coalville)	81,794.22
Salt	29,662.82
Castings (Salt Lake foundries)	12,867.10
Beef and vegetables	21,724.03
Hauling and sampling ore	55,853.10
Sundries, powder, oil, machinery, candles, groceries, N. Y. & S. F. offices	310,323.54

ONTARIO DIVIDEND NO. 1 TO 197.

1877.	No. 1 to 18	$ 900,000
1878.	No. 19 to 39	1,050,000
1879.	No. 40 to 51	600,000
1880.	No. 52 to 63	600,000
1881.	No. 64 to 75	875,000
1882.	No. 76 to 87	900,000
1883.	No. 88 to 90	225,000
1884.	No. 91 to 102	900,000
1885.	No. 103 to 115	975,000
1886.	No. 116 to 127	900,000
1887.	No. 128 to 139	900,000
1888.	No. 140 to 151	900,000
1889.	No. 152 to 163	900,000
1890.	No. 164 to 175	900,000
1891.	No. 176 to 187	900,000
1892.	No. 188 to 197	750,000
Total		$13,175,000

The Ontario has recently passed a regular dividend, a circumstance which caused quite a flurry in financial circles for a while. It was not caused by failure of the ore supply, but the enormous expense of running a drainage tunnel.

TOTAL PRODUCT.

The total output of the Ontario from the starting of the new mill February 1, 1877, to the end of 1892—sixteen years—was 414,405 tons (dry) of ore, out of which was obtained 27,876,469.45 ounces of fine silver.

The company has been preparing to change from the use of wood in roasting ores to manufacturing gas from Rock Springs coal to do the work. The apparatus is already in and only awaits a short stoppage to make connections. It takes eighteen cords of wood per day at a cost of $5.75 per cord, while it is estimated that ten tons of coal at $4.75 per ton will do the work better and with less men to operate, making a saving of fully $60 per day. The Marsac mill has been operating about two years with the gas and finds the saving proportionately as large.

The excess of disbursements above receipts is drawn, of course, from surplus account.

For a number of years the water has been drawn from the Ontario sixth level by a drain tunnel about 6,000 feet in length. In 1888 a drain tunnel was started from the east or Provo side, three miles distant, to drain the mine to the 15th level. February 15th, 1891, this tunnel was in 6,170 feet, and the Ontario Shaft No. 2 was down to the 15th level, where the tunnel is to intersect it. The mine has still a vast amount of opened but unstoped ground above the 10th level. The selling price of the shares is from $40 to $44; there are 150,000 shares, par $100; holders have come to repose trust in them as if they were United States bonds. The mine has passed the monthly dividend of 50 cents a share but about six months, when No. 2 hoist burnt down, in fourteen years. No one familiar with it doubts that this will continue 15 or 20 years longer.

The Daly mine is a great shipper. Sales of the proceeds of ore bring the company immense sums, a large proportion of which are paid in dividends.

The Anchor property is a group of claims 1,200 feet in width by 12,538 feet long, beginning near the west end of the Daly and running west, comprising the old Utah and White Pine properties, and many other contiguous claims. Mention has been made of a trial shaft sunk to the 600-foot level, and of the large vein it developed. It developed also so much water that a drain tunnel was driven from a point 6,600 feet down the gulch and low enough to intersect the shaft on the 1,200-foot level. Pending work to connect shaft and tunnel, the hoisting works caught fire and were destroyed. New works have been erected, and meantime the company extracted ore from the tunnel level in large quantity. The vein is from 18 to 70 feet in thickness, dips to the northwest, and strikes northeast and southwest. Large chambers have been cut out, the openings exposing more ore.

The Daly West ground is a group of claims about 2,000 feet wide by 3,000 feet long, joining the Daly on the west, and the Anchor ground in part on the north. It belongs to the owners of the Ontario and the Daly, and can be cheaply drained and exploited and laid off properly for ore extraction through or by means of the workings in those mines when the owners are ready to do so. It is supposed to be as good ground as either of them.

The Woodside Company owns eight claims in Woodside Gulch, out of which they took $444,000 in 1889, and shipped 161,880 pounds of ore in 1891.

The Mayflower comprises a group of claims near the Woodside. The owners extracted 1,560 tons of ore in 1889, and 2,629 tons in 1890, when they were enjoined, pending decision of some question of title, and obliged to close down. They have started up again, however, having shipped 8,655,470 pounds of ore in 1891.

The Massachusetts (old Empire), comprising twenty claims, lies about one mile west of the original Ontario ground. It is well equipped. A fork, at least, of the Ontario vein is believed to run through this ground.

The Alliance (old Sampson) is a group of eight claims, immediately east of Pinyon Hill, on a line west with the original Ontario and Massachusetts; the Daly and Anchor diverge to the southwest. It is at the head of Webster and Walker Gulch, and much higher in altitude than the Massachusetts. The vein is in limestone, is fifteen feet thick, and strikes through Pinyon Hill southwesterly a mile or more, the Crescent and the Apex each owning a part of it. The working shaft is intersected by the Hanauer tunnel at a depth of 520 feet. This tunnel has been extended beyond the shaft on the vein (as a level) to the end line of the property, and to a connection with the Crescent working incline for the benefit of the Crescent Company. Cross-cutting and running west in the vein shows the vein to vary in width from twenty to forty feet; to pitch about fifteen degrees from the vertical; to be in limestone still; and to be filled with quartz, brecciated lime, clay, talc, iron and manganese, with occasional bunches of ore.

The Crescent property comprises about ninety acres, the ore occurring in a channel twenty rods wide and a quarter of a mile long, falling off to the northwest with the face of the hill and about 100 feet below the surface. This ore sheet crops out in the eastern face of Pinyon Hill, which is on that side a ledge about 400 feet high, and their veins or fissures come occasionally to the surface from the ore body on the northwestern slope of the hill. More than 100,000 tons of ore have been extracted and sold from this ore body.

The Sampson (or Alliance) vein cuts across the head of the property, and is reached, at a depth of about 400 feet, by a 1700-foot tunnel run in from the northwest. Upon this vein, which is here 50 feet wide, at the inner end of the tunnel, machinery has been placed, and a working incline sunk to intersect the Hanauer tunnel, 400 feet. This will save pumping and give 400 feet of dry stoping back. The ore in this vein is scattered; 2½ tons are concentrated into one. The company have a concentrating and sampling mill; 5 miles of tramway between mine and mill, with an average grade of 400 feet per mile; boarding and lodging houses, etc. The property is regarded as in better condition than ever before.

The Apex is a companion property to the Crescent, lying south and contiguous, comprising about twenty claims, and covering a part of the blanket ore-bearing formation of the Crescent. Large bodies of low grade ore have been exposed by the extensive workings. About 199,230 pounds were marketed in 1891. This ore is from the Sampson vein, which is cut at a considerable depth by a tunnel from the side of the hill facing Thayne's Cañon. This tunnel continues to a connection with the Crescent workings. The workings in the Apex extend 200 feet below this tunnel. Work seems to have been confined to exploiting.

Other paying mines are the Northland-Nevada, opened in June, 1889; the Creole, lying on the hillside above Park City; the Deer Valley Consolidated, 13 claims northeast of the original Ontario location; the Constellation, a group of 5 claims northeast of the Ontario; the Golden Eagle group of 5 claims, joining the Constellation; the Whitehead group of 24 claims, east of the Constellation; the Putnam group of 12 claims, near and partly between the Daly and Anchor; the Morgan group, near the Anchor, 13 claims, has been in the courts, and ·all questions of title have been settled and the property incorporated as the Meears Consolidated; the Roaring Lion, adjoining the Crescent, is a strong vein of rich ore similar to the best ore produced by the Crescent; the Jupiter property, 13 claims, mainly patented, at the head of Thayne's Cañon, about a mile southwest of the Crescent; the Silver Key, 4 claims, south of the Apex; the Silver King, on the Woodside and Mayflower vein; the New York group of 5 patented claims; the Lucky Bill, half a mile south of the Daly; the Comstock, 4 claims, on the opposite side of Thayne's Cañon from the Crescent; the ground contains a large vein much like that of the Crescent; the Gem, 4 claims, adjoins the Comstock; the Steele group of 6 claims, a mile below the depots.

Many mining properties have not produced much ore, but remain to be mentioned; amongst them the Dolberg group, the west Ontario group, the Black Diamond and Nimrod, the Rosebud group, the Reed group, the Kerr group, the Hoyt group, the Park City group, the Lundin and Anderson, the Rosecamp and Glen group, the Denhuff group, the Creole No. 2, the Typo group, the Kentucky group, and the Hughes and Bogan groups.

Besides these, there are hundreds of promising prospects scattered all over these hills, from the Cottonwoods to Provo River, and from Deer Valley nearly to Midway, a district containing fifty square miles. The country is wet and the drift heavy, making the development of prospects into paying mines slow work. Very little capital on the outside has ever gone into the district. It has had to depend upon its output for the means of increasing its output. Yet it may be truly said that there is no district in the entire mining section which offers greater inducements to capital to engage in mining than this.

The Union Concentrator, built in 1889, at the junction of Empire and Woodside Gulches, has a capacity, running day and night, of 120 tons in 24 hours. The machinery used is Dodge stone-breaker, Wall rolls, Cornish rolls, jigs and tables. It is heated by steam from the boilers, and can be run winter and summer.

The Park City Sampling Mill, owned by Richard Mackintosh, of Salt Lake, does a large business constantly.

The shipments by rail for 1891 were:

	Pounds.
Ontario	24,803,410
Daly	10,882,520
Anchor	19,736,660
May-Flower	8,655,470
Crescent	533,230
Vareo's Con	690,490
Woodside	161,880
Nevada-Northland	178,960
Apex	199,230
Creole	222,700

WASATCH COUNTY.—Blue Ledge District lies on the eastern slope of the divide between the Provo and the Weber, and is in Wasatch County. The Glencoe is at present the leading or most promising mine in the district. It consists of a group of six or eight claims. In the old workings there was a strong vein almost unbroken for 300 feet, but rather low grade and carrying too much zinc. An adit tunnel, 150 feet below these old workings, is approaching the old ore body, and is full size in ore of a much better quality and carrying less zinc than the old ore. The mines are about two miles a little south of east of the Ontario.

There is valuable property in McHenry Gulch, to-wit: the Wilson & Barrett, the Lowell, the McHenry, the Hawkeye, the Boulder, and southward of the gulch the Free Silver, the Wasatch, and many others. All these are groups of from two to a dozen claims, and on some of them much heavy and expensive development work has been done.

Good judges do not doubt the existence of great mines on the east side, as it is called, but the ground is broken and thrown by eruptive dikes, and ore in paying bodies, if it exists, probably lies deep.

UTAH COUNTY.—A spur of the Wasatch striking eastward and then northward forms the line in this locality between Summit, Wasatch, and Utah Counties, and also between Uintah, Snake Creek, and Blue Ledge districts, the latter in Wasatch, Snake Creek in Utah County. Headquarters of all these districts, it will be understood, is Park City. The mines in Snake Creek are about 8 miles from Park City over a high divide, yet it is the best way out at present.

The Southern Tier has been opened to a great depth and some shipments have been made. Amongst other groups of claims upon which considerable work has been done are the Newell, the Steamboat, and the Levigneur claims. The formation is mixed and pretty badly broken and tumbled up on the surface. Nevertheless the miners are developing regular and continuous veins, which produce very good ore. There are copper lodes and ledges of marble; and at Midway, on the Provo River, in plain sight from the mines and not far away, there are hot springs and quite an area of the "formation" which in many places these hot springs deposit. With a railroad on the Provo River, Snake Creek will be heard of to some purpose in the mining world.

A GREAT MINING FIELD.—Before completing the review of Utah County mines, the reader may as well return to Salt Lake City, and take a general glimpse of the field. About thirty miles east of Salt Lake City, the counties of Salt Lake, Utah, Wasatch and Summit corner at the apex of Clayton's Peak, in the heart of one of Utah's great mining fields. This field is from fifty to one hundred square miles in area; its absolute altitude is from 7,000 to 11,000 feet, and it is extremely rugged. It is here that the Wasatch range is the highest and most massive. The formation is quartzite and lime, held up on granite shoulders. Much of it has been gouged and worn away by erosive agencies. From its culmination the water flows in all directions. American Fork, Little and Big Cottonwood, and Mill Creek westward, and small unnamed streams south, north and east, into the Provo and the Weber. The western limb of it is accessible only via the streams named from Salt Lake Valley. The eastern limb is reached by the Union Pacific from Echo on the Weber. A rail and tramway run to Alta at the head of Little Cottonwood from Brigham Junction, which is ten miles south of Salt Lake City. A good wagon road runs up American Fork, starting from the town of American Fork, which is about thirty miles south of Salt Lake City, to the Miller mine, probably 11,000 feet above the sea; and a wagon road also runs up Big Cottonwood to the Lakes, and crossing the divide down Thayne's Cañon to Parley's Park and Park City. The mines pay no attention to divides, although these are the boundary lines of counties and mining districts.

The palmy days of the Cottonwoods and of American Fork passed away a dozen years or more ago, with the exhaustion of the surface bonanzas of such famous mines as the Emma, Flagstaff, Joab Lawrence, Miller, Prince of Wales, Reed & Benson, etc. Work has never ceased altogether, however, although it has ceased on hundreds of prospects, and on scores of mines. This is due to the same incidents that everywhere embarrass mining — lack of means being the principal. It takes a mine to make a mine, Spaniards say, and it is true. Some of our greatest mines would be as dead and unknown as any of the 1,500 patented mines of Utah had they not at an early stage passed into the hands of men of ample means; men able to put in a good deal of money bef. re they took any out. There ought to be 200 producing mines on the Cottonwoods and American Fork, and some day there will be. Some accidental strike will recall attention to this mining ground, so accessible from the valley; men will again flock in there; work will be resumed on properties partially developed by men full of pluck and with means, and also on the merest prospect holes; and more money will, in the future, come down these streams in a year than is taken at present from all the mines of Utah. Fifty mines might be named in the district that need nothing but work, exploitation, to become profitable producers. And there are four times fifty more, probably equally meritorious, which were never worked enough to be known. There are about a dozen which are worked in a small way, and send out a little ore every season.

On the eastern side of the field a rich company early became engaged in mining, and so there has been no abandonment, although this company's was for years the only productive property in Summit County. The Ontario mine was discovered by the merest accident, the turning of a loose cobble-stone of ore in the bushes on the side of Ontario Gulch. A narrow little trench a few feet long was found to be full of rich ore, and the "find" was sold to Hearst, Chambers & Haggin for $30,000. The Ontario Silver Mining Company was organized and a great deal of money expended in mill and mining plant and development before any ore of consequence was taken out. The reader of these pages has already some idea of what has been done since. If, as we are assured by the superintendent, the mill has three more years' work above the tenth level, it will have been seventeen years exhausting the mine to that level. At the same rate, with the long drain tunnel completed and taking the water from the fifteenth level, there are eight and a half years' work between the tenth and the fifteenth levels; and, if the formation continues and the vein retains sufficient fertility, it may be worked by pumps to the twenty-fifth level, seventeen years more, or in all forty-two and a half years. Dividends of $900,000 a year have been so long paid, that, as has been said, they are looked for as confidently as the payment of interest on Government bonds. Forty years of life for such a mine means the wresting from that fissure of $75,000,000, and the payment of $40,000,000 in dividends. Yet one year the ore ran down to $67 per ton, and one-fourth of the mine was offered for $375,000, and after examination declined. And yet again, notwithstanding the fact that the first ore taken from the little trench spoken of sold in Salt Lake for $245 a ton, the chances are ten to one that if the owners had not had unlimited means, this unequaled mine, which was naturally a water geyser, would have been abandoned the same as the Davenport or the Wellington or the McHenry or the Hawkeye or the Lowell are, and as the Crescent, the Woodside, the Wasatch, and a hundred other Utah mines at one time or another have been.

The Daly is a continuation of the Ontario westward, and it took four years of outlay to work this mine up to the dividend-paying stage. Blind tunnels were run into the banks of the gulches, and a shaft put down 500 feet, and levels and cross-drifts run, pumps set and compressors and hoisting plant put on, and a mill built; and long after that, when dividends had begun, a considerable interest in it was offered for sale at the rate of $200,000 for the whole. Its total dividends are now nearly ten times $200,000, and its life bids fair to extend side by side with that of its foster father, the Ontario.

There are mines still west of the Daly and east of the Ontario, and alongside of both, doubtless as good as they are. It is a wonderful district, full of prospect holes, of tunnels and adits and shafts stopped just short of fruition. There was the Woodside, abandoned for eight years, then taken up and proved a bonanza, and that has revived a whole group of mines in the vicinity, and in other localities, and thoroughly broken up the superstition that there was but one mine or ore vein in the district. The Anchor, the Alliance, the Crescent, the Apex, and at least a score of groups within three miles of Park City, need nothing but judicious working to make great mines of them.

AMERICAN FORK.—With all the other mining districts of Utah, American Fork has experienced a resurrection, though but few mines are being worked, and their shipments and profits, if any, are not reported. They are "all there," and are bound to come to the front.

SALT LAKE COUNTY.—LITTLE COTTONWOOD —The mines of Salt Lake County are at Bingham Cañon in the Oquirrh, and on the Cottonwoods in the Wasatch, both connected with the Jordan smelters and with Salt Lake City by rail and tramway.

Concerning the mines of Little Cottonwood, a writer in the New Year's Salt Lake *Tribune* furnishes the following review for the year 1890, which is measurably applicable to last and this year:

"While the ore product the past season was not much in excess of the last few years, it was sufficient to greatly encourage the few who have had the courage to remain, satisfied that time would eventually demonstrate that their confidence in the permanency of the ore body was well founded.

"The developments in the Emma have been more encouraging than at any time in the past ten years. While no very large bodies of ore have been encountered, a number of promising streaks have been found, yielding a good grade of ore, running up to 200 ounces, and these streaks have widened out as depth is obtained, and promise to increase the product very materially during the coming season. While developing these richer veins, they have struck a good body of concentrating ore, which, in all probability, will alone pay all the expenses of operation as soon as the facilities for landing the ore at the concentrator are perfected.

"The Flagstaff for the first time in ten years is again in ore. During the latter part of the season they encountered quite extensive bodies of mineral. On the first of November this mine had 500 tons of ore in the bins ready to be sent to the samplers. Their ore body has widened out to such an extent that they are taking out about 100 tons a week.

"The City Rock has not been worked the past season by the owners. It has been in the hands of leasers during the greater part of the year, who have made a very good showing. Just at the close of the season they struck quite a body of ore. The prospect is very good for this property.

"The Vallejo is not worked very energetically. The ore is of a very high grade, and when the pockets have frequently been exhausted others have always shown up as development proceeded. The Vallejo has undoubtedly large bodies of ore remaining awaiting development.

"The Toledo, at one time a mine of great merit, has been operated for the past few years by leasers who have been handicapped by the great depth of water in the lower levels, which they have inadequate facilities for handling.

"The owners of this valuable property some years since built a compressor for the purpose of furnishing air and draining the water from the mine. The compressor was unfortunately totally destroyed by a snowslide, which so disheartened the owners that they have never resumed operations. Yet it is a well known fact that good bodies of high grade ore exist in the lower portion of the mine. The Hoboken is an old location which has taken on a new lease of life during the past season, and promises to become a good producer in the near future.

"The Golconda, immediately south of the Vallejo, was a great producer in former years, but as soon as the first body of ore was exhausted, work was given up by the owners. The mine has been in the hands of leasers for several years past. From the success they have met with it is certain that if vigorously worked it would still be a valuable property.

"The Montezuma group, comprising the Savage, Hiawatha, Montezuma and other properties, is owned by an Eastern syndicate. Operations on this group have been retarded for several years, awaiting the completion of a long tunnel and upraise, which has at length been completed. Under the supervision of Mr. Thomas Buzzo, one of the most competent mining superintendents in the country, the property is at last in condition to be workrd systematically. Considerable ore is in sight, and if Mr. Buzzo's experiments are realized a great quantity of ore will be sent to the front in 1891.

"The Jack mines, owned by the genial Charles Sickler, and his friend, James Tainsch, is looking well—from the appearance of the ore that has been exposed; it looks as if the owners would be able to spend the rest of their days in merited comfort.

"The Toledo Dump, under the management of Fritz Rettisch, has made several shipments of concentrates during the summer past.

"The Highland Chief has been in the hands of leasers for several years. If Mr. Chisholm would put a little of the great wealth he is deriving from Tintic into this property it would doubtless resume its former productiveness.

• " These are the principal mines from which shipments have been made during the season of 1890. A number of new prospects have been opened up, which are very encouraging. The King mine has been worked and incorporated during the season just ended. An able mining expert examined this property and made a very flattering report, on the strength of which the mine was immediately incorporated. Active operations will commence as soon as the weather will permit, and from present indications another valuable property will be added to the long list of Utah ore producers.

" The ore shipments and values are as follows:

MINE.	TONS.	AVERAGE.
New Emma........	250	110 ozs.
City Rock...............	175	72 "
Vallejo........	150	160 "
Toledo........	60	125 "
Hoboken	70	65 "
Pittsburgh	60	32 "
Montezuma	150	90 "
Jack.....	10	225 "
Toledo Dump	25	110 "
Highland Chief................................	30	70 "
Total....	980	

Flagstaff, large quantity on hand, averages 40 ozs. No shipments. ·

" Great credit is due the English owners of the Emma and Flagstaff properties. They have sustained the camp during its past seasons of depression; the Emma, to a far greater extent than the Flagstaff, having spent hundreds of thousands of dollars in their persistent endeavors to find the lost ore bodies, and now that their efforts bid fair to be crowned with success none should envy their good fortune. Mr. Henry Clay Wallace has been the efficient manager of the Emma for the last three years. If the mine comes to the front he should have all the credit, as he has worked night and day in his desire to succeed.

The Greely mine is a mile from the tramway, near Alta. A little work is going on in the Peruvian, the Oxford and Geneva, and, indeed, in several other once noted mines.

BIG COTTONWOOD.—This district has a number of good mines, while there are many prospects which would become paying properties if sufficient capital and energy were applied to their development. There were more signs of activity the past few months than for several years.

The Maxfield is the leading mine. About the middle of 1888 the ore in the mine was cut off by a fault, and it took two years to find it again and make the necessary connections to resume the extraction of ore. They run drifts during this period aggregating 6000 feet, and found the new ore body 175 feet southwest and 250 feet above their old workings.

Two dividends of $9,000 each have been paid. Mr. W. F. James, the manager, is pleased with the condition and prospects of the property.

The Congo, a mile and a half above the Maxfield, is an old property which laid idle for years awaiting the settlement of the estate of Dr. Norton, the former owner. It passed into the hands of a company who are running a tunnel and opening up a good property. This tunnel has been driven 200 feet.

The Queen Bess has long been tied up, but has been purchased by parties who have fair promise of making it a good mine. Mr. Baker shipped a few tons of ore from his property, and developments are quite favorable. Some, but not much, mining was done on Kesler's Peak, on the Prince of Wales, about the lakes, and on Scott's Peak.

The Gipsey-Blair property was sold in 1890. The new owners put in new machinery and are doing some work.

The Reed & Benson, in early times, turned out $300,000, but the ore pipes, which were followed, made such labyrinthine workings that, at a depth of about 500 feet, work ceased in the mine, and a tunnel was driven 2,200 feet on a level 500 feet below the lowest of the old workings.

BINGHAM CAÑON.—We are now through with our review of the great mineral field east of Salt Lake City in the tops of the Wasatch. The scene is transferred to the Oquirrh Range west of Jordan Valley, or to that part of it known either as Bingham Cañon or West Mountain District, being so much of the eastern face of the range as has been cut into a fan-shaped series of ravines and ridges by the melting snows which find their way out through Bingham Creek. The town of Bingham, strung along the gulch at the entrance of Carr Fork, where the gorge is deepest, is about 26 miles southwest of Salt Lake City by rail. In the seventies the bed and sides of the gulch were burrowed, tunneled and sluiced in many places as gold placers, and the end of gold placer mining in the gulch has not yet come, if, indeed, the beginning has. The great stream of lead-silver ores which has flowed out of the cañon to the Jordan smelters for 20 years, and which is now swelling in volume, will be our first theme. In the earlier years the output was extraordinary, but when the oxidized ores of the surface had been mainly used up, the output fell off to a point much below what it is at present, and for the past few years it has steadily increased. The great ore channel of the district strikes northeasterly from the summit of the range (Oquirrh) about three miles to the valley, crossing upper Bingham, Bear, Yosemite and Copper Gulches. Below Bear Gulch its course is cut off from the range by Bingham Cañon; the exposure is to the south, and the ground is comparatively dry. The Brooklyn, the Yosemite, the Yosemite No. 2, the Miner's Dream, the Wasatch and the Lead mines are on this part of the great ledge or zone.

(This review has special reference to 1890, but with the necessary substitutions, is fairly applicable to the subsequent time.)

The Brooklyn comprises several locations adjoining the old Telegraph on the northeast. The hoisting works and concentrating mill are in Yosemite Gulch, 300 to 400 feet lower than the divide between it and Bear Gulch. The main incline is on the quartzite footwall, and pitches northwesterly at an angle of 45 degrees. Fifteen levels have been opened along the foot, aggregating in length about four miles. The ore makes in pipes or chimneys 100 to 150 feet long on the course of the ledge, and from 2 to 12 and 20 feet thick. These chimneys or pipes go down with slight change on the whole, either in dimensions or character of contents. The ores are galena, carbonates, and sulphates, 60 per cent requiring concentration to bring it to shipping grade, to-wit, 10 ounces silver and 50 per cent lead. Concentration is by jigs and tables, and costs 75 cents to $1 per ton. Three hundred to 600 tons are shipped per month by the Brooklyn, and have been the past seven years. The vein is regular and well defined on the footwall side. The hanging wall, a lime shale, is much less easily located, and is believed to be 400 feet from the footwall.

The formation is complicated by the existence of the Yosemite, comprising several locations, on a vein very like the Brooklyn, parallel, pitching perhaps 20 per cent less, and 400 to 500 feet toward the hanging country, on which the workings are extensive both in depth and lineally, although less extensive than the workings of the Brooklyn. The works are in Yosemite Gulch; the concentrating mill three miles below, in Butterfield Cañon. The ground rises each way from Yosemite Gulch, and a great deal of it is still unexploited. Water was struck in the working incline on the 6th level, and the ore became pyritous, but down near the 8th level it changed to galena, carrying 16 ounces and upwards of silver. The Brooklyn also struck water in one of its ore pipes on the 12th level, but the lean iron pyrites which came in thereupon gave place to galena between the 13th and 14th levels. The Brooklyn and Yosemite are now owned by the Lead Co., and their ores ore run down to the Lead Concentrating Mill on the railway, near the mouth of the cañon, over a gravity tramway about 5 miles in length.

In Copper Gulch, half a mile further east, and 200 or 300 feet lower, these two veins are known and worked as the Lead and the Yosemite No. 2, and as the Wasatch and the Miner's Dream, respectively. They have the same general characteristics and yield the same kind and quality of ores as the Yosemite and the Brooklyn. The Wasatch and the Miner's Dream are opened by an incline to the depth of 600 feet; Yosemite No. 2, by a shaft to the thirteenth level.

West of the Brooklyn and the Yosemite, in the Old Telegraph on Bear Gulch, where the exposure is to the northward, these two veins, if such they are, seem to have become one. The clean marketable ore on this property, which is a consolidation of twenty-one locations, reached, in places, a width of nearly 200 feet, and the lean iron pyrites upon which the oxidized ores bottomed at the level of the bed of the gulch—here and above the true water level—is estimated at four millions to five millions of tons. Out of the ridges bordering Bear Gulch, more than 70,000 tons of oxidized ores, which sold for upwards of $1,500,000, have been taken.

Still west of the Old Telegraph, in the Spanish, the mineralized zone is 600 feet wide, the ore making in pipes and kidneys of all shapes and dimensions, but with a certain regularity of strike and dip. On the surface there was a vast body of oxidized ores.

The Jordan lies next west. At its intersection with the South Galena and the Utah the oxidized ores of the surface worked out over a hundred thousand tons, worth more than $2,000,000, and there now lies in the same vicinity a million tons of $20 quartz, in which gold and silver are so combined that no way has yet been found to work it without a loss of most of the one or the other metal. Four hundred thousand tons of similar material, bearing $8 and upwards per ton in gold and about the same in silver, constituted a hillside above the bed of Carr Fork on the Stewart property, half or three-fourths of a mile north of the Jordan, believed by competent geologists to be part of the same deposit. On all this upper part of the mineral belt the snowfall is heavy, it melts slowly, sinking instead of running off, and the ground appears to be full of water clear up to the surface drainage. At all events, the surface drainage is the line of division between oxidized and base ores.

The Jordan, the Spanish and the Old Telegraph were paralyzed for years by the exhaustion of their oxidized ores; but as methods have improved, work has been resumed, and their output is yearly increasing, shipments comprising remnants of surface carbonates, generally requiring concentration, and galena, more or less mixed with iron pyrites, which has to be roasted and in much of it the pyrites dressed out. All these mines have concentrating mills, in which, by a careful adjustment of jigs, screens and tables, determined or regulated by experimenting, galena and iron pyrites are obtained as separate products, cheaply and without great loss. The latter generally carries a fair proportion of the silver and has a value as fluxing material.

As has been stated, the deepest workings in the Brooklyn and the Yosemite seem to indicate that the pyritous zone is less than 200 feet thick, galena predominating below, a galena twice as rich in silver as the surface carbonates and sulphates. If this prove to be the fact, it will lead to deeper workings on the upper part of the belt, where the ores appear to be in practically unlimited quantity. But if the pyrites persist to the deep, the future of the district must mainly depend upon the utilization in some manner of low grade pyritous ores. To accomplish this, cheap and perfect ore dressing, saving of all the contents of value, cheaper transportation, cheaper fuel, and cheaper labor than are available at present, are indispensable conditions.

A great number of locations have been and are constantly being made.

A number of groups of claims in upper Bingham, purchased by the Niagara Company, organized and managed by Mr. P. A. H. Franklin, include the following: The Indiana, the Miller, Idaho, Accident, Silver Plume, Red Cloud, Dead Thing, the Utah group of five claims, the Spanish, Black Hawk, Bonnie Blue Flag, Murphy, Crescent, Canby,

Climax, Ajax, Defiance, Union, Lady Franklin, Quaker City, Live Pine, St. Marks, Mack S., Alameda, Austin Ray, Red Cap, Henrietta, Red Warrior, Portland, Sturgis, Safe Guard, Rupert, Oquille, Dartmouth, Bullion, Ben Bolt, Niagara, Palon, Dickerman, Ohio, and perhaps half a dozen more. Many of these claims have given up great quantities of ore, and in many of them there are large bodies of ore in sight. Old openings have been cleared out and retimbered, and new works begun, notably a new working shaft in the heart of the ground, and a tunnel for drainage and working purposes under-running the property for half a mile, from 350 to 1,200 feet below the surface. They have a concentrating mill, capacity 120 tons per day, and a very large boarding and lodging house and other conveniences. In purchasing these mines and initiating the new work, $300,000 to $400,000 was expended. Altogether it is a vast mining property, containing, no doubt, millions of tons of ores that, with means and skill and pluck, may be profitably extracted and reduced. But the company may have to put in a good deal more money in preparatory work before they can take out the ores to the best advantage, and it will not do for them to cut off the supply because of every flurry in the stock exchanges. This enterprise is business, not speculation. It can be made, according to its management, one of the most profitable mines in the world, and of very great benefit to the district and the Territory, or one of the most noted failures amongst mining enterprises.

Amongst other mines in Bingham, which are now worked under lease or by their owners, are the Old Telegraph, the Jordan, the South Galena, the Winamuck & Dixon, the Buckeye, the Lucky Boy, the Silver Gauntlet, the Neptune, the Live Yankee, the Monitor, the Highland, the York, the Petro, the Minnie, the Leonard, the Agnes, the Pisa, the Mary, the Morning Star, the Last Chance, the Frisco, the Nast, the Stewarts 1 and 2, the Big Giant, the Little Cottonwood, the Sampson, etc. It is not worth while to try to give an idea of the amount and nature of the openings on these mines, or of their condition and prospects. Output is all that counts in mining.

Following is a list of the mines which shipped 100 tons of ore or more in 1890, namely:

Mine.	Tons
South Galena	9,620
Brooklyn	8,092
Yosemite No. 2	2,610
Old Telegraph	2,500
Spanish	2,100
Niagara	1,500
Lead and Yosemite No. 1	1,396
Utah	1,210
Winamuck	715
York	650
Highland	620
Dixon	310
Rough and Ready	212
Silver Hill	138
Markham	137
Silver Gauntlet	133
Buckeye	133
Silver Shield	114
Last Chance	107
Fire Clay	102
Total	32,399

Add 1,423 tons shipped by 40 other mines combined, and we have 33,822 tons as the amount of ore shipped from Bingham in 1890.

The output for the year 1891, reached the enormous figure of 138,475,789 pounds, more than double that of the previous year.

The mines of the district seem, in general, to be steadily improving, both in product and promise. None of them have been explored to any depth below water level. Most of them are worked by lessees, depend upon their product for development, and even for plant, and are necessarily worked with the greatest care and economy. Could this district—and this is equally true of all our mining districts—command means by assessment to outfit and open their mines systematically, as the Comstock mines could and did for twenty years, Utah mining would enter upon a new era, and our output would be doubled twice over.

TOOELE COUNTY, RUSH VALLEY DISTRICT.—A stub railway, part of the Union Pacific system, runs from Salt Lake City west, passing round the end of the Oquirrh Range via the lake shore, and bearing southward to within a mile or two of Stockton, so far the only mining town of Tooele County. It is 10 or 12 miles south of Great Salt Lake, and about 40 miles from Salt Lake City. The mineral belt, beginning at Stockton, strikes southerly along the foothills of the western slope of the Oquirrh Range, a little diagonally with the range itself, throwing it up toward the summit further south, as at Dry Cañon, Ophir and Lewiston. The belt is a mile or more in width. There appear to be two systems of veins at Stockton, one striking east and west, in which the main ore bodies make; the other north and south, thinner, less persistent and apparently feeders. The formation is quartzite and lime, underlaid by syenite. Granitic porphyry dikes cross and disturb the veins. The gangue is oxide of iron, quartz, spath and clay. The ore is galena and carbonates, free from base metals, and very desirable as a flux for dryer ores. The ore makes in well-defined pipes or chimneys, of which there may be five or six in the course of a thousand linear feet. The water level is 700 or 800 feet below the surface. None of the mines appear to have gone below it as yet.

Several incorporated companies and sundry individual miners are working and developing more or less promising properties about Stockton, and the business and its returns are steadily increasing. Most of these mines are worked by lessees. The Honerine is, perhaps, the leading mine at Stockton. The mine is a bedded vein in magnesian limestone, crossed by dikes of porphyry and a series of thin fissure veins. It is equipped with steam hoist and is opened to the water level, about 800 feet, by working incline and levels 100 feet apart.

(This account was not, excepting the concluding part, written for the present year, but is mainly applicable.)

DRY CAÑON.—At Dry Cañon lessees and part owners have been extracting ore of good quality from a group of claims consisting of the Brooklyn, Elgin, Belfast and Trade Wind. The Mono turned out a good deal of exceedingly rich ore from its surface bonanza, but it has long laid idle. The owners of the Hoistead have a valuable property.

OPHIR.—The Ophir Hill Mining Company own the Miners' Delight, literally a mountain of low-grade ore, to concentrate which they have completed a mill capable of handling 150 tons a day. A hoist is operated by compressed air from the mill, which is 650 feet below the mill. The air is carried up in pipes, and the ore dropped down on a tramway a distance of 2,300 feet. The mine is an old one; it has long been worked under lease, and it is not in very good shape. Righted up, it is expected to last the life of a generation.

The Utah Gem is a contact between lime and slate shale about 12 feet thick, fed by a series of stringers from the footwall country. The ore makes in pipes, and can be selected to a very high grade. Mr. L. E. Holden, the owner, has a 10-stamp mill, and has done considerable marketing.

The Monarch and Northern Light, long dormant on account of litigation, are regular shippers of high-grade ore. The vein is large, dips to southwest about 30 degrees, and is opened to a depth of over 900 feet. The ore in the Monarch is a milling ore; in the Northern Light, a lead carbonate containing chloride.

The Buckhorn group is in the hands of a company, and is systematically worked.

Besides these properties there are the North Star, with a 12-foot vein of ore; the Gladstone, the Chance, the Forest group, all eligibly located; and scores and hundreds of others await the application of capital to make them contributors to the wealth of the whole country.

Mr. Frank Burk, in a letter in the Salt Lake *Tribune*, voices the chief need of this region as follows:

"The lack of railway transportation is the only drawback to this district (Ophir), and south along the Range to Boulder and North Tintic. If the Deep Creek Railroad people would send a competent engineer to Stockton and Ophir, with instructions to examine the feasibility of a route for their road by way of these points and south along the base of the Oquirrh Range, through either Boulder or Twelve Mile Pass, to Eureka, Silver City, and then due west through West Tintic, Desert, Death Cañon and the Dugway district to Clifton and its destination, Deep Creek, they would find that such an engineer would report ore in great quantities on nearly every mile along the route. The object of the road is to reach the mines on the western border of Utah, and beyond, in Central and Southern Nevada. If it is an object to reach out after this ore, 200, 300, 400 miles away, why is it not an object to reach the mines so much nearer at hand? There is no better field for investment of capital in the entire west than that which the Oquirrh Range offers, on both slopes, all within 100 miles of Salt Lake City. The mines will remain undeveloped and dead property, as for the most part they are and long have been, until money can be obtained to open them. The money will come with the railway, and it will not go far in advance of the railway after low grade ores."

The Union Pacific should extend the Stockton line down the west base of the Oquirrh to the vicinity of Tintic, there connect their Tintic line with the Stockton line, and go west, on the route Mr. Burk advises. Should they do so they would have all the business between Salt Lake City and Deep Creek within a year or two that a first-class line could do. Beyond Deep Creek, the Union Pacific Company is aware, from its own investigation, that a railway would have all it could do, almost from the day of its opening. It is the stretch between Salt Lake City and Nevada of which the company is doubtful. The writer agrees with Mr. Burk that the capacity of this stretch to develop railroad business, if the route he suggested be taken, has been very much underestimated, even by the most enthusiastic devotees of a railway to the Deep Creek country.

The great Utah mine, of Fish Springs, is already a bonanza, being as extensive a shipper as is consistent with the high price of teaming. As a producer of rich ore it rivals the Ontario. There are many others here and at Dugway, in the same region of country, which would be wealth-producers with a railroad.

PIUTE COUNTY: MARYSVALE.—This is the mining town of Piute County. It is seventeen miles above Monroe, on the Sevier, ninety miles from Juab, the nearest railway station. This was one of the first mining districts organized in Utah, and eighteen years ago wagons loaded with Marysvale ores were not an infrequent sight in the streets of Salt Lake City. The haul of 200 miles, however, proved too expensive, and later the haul of ninety miles.

The following, from the *Christmas Herald*, 1891, is a very complete showing:

"Marysvale is located in Piute County, just over the divide from Sevier County. It is about nine miles west of a direct line south of Salt Lake and about two hundred miles south. It lies in a little valley that is protected on all sides by the hills which form a barrier against heat or cold. The low pasture lands are covered with thickly growing hay and marsh rushes, that form a curious contrast to the mineralized hills surrounding them. There is salt, alum and fire clays, besides the rich mines of the Bullionville, or Ohio, and Mount Baldy districts, situated on the west and south of Marysvale proper. Marysvale was a mining camp as early as 1862, when a party of Mormons came in and settled on the fertile farm lands. The Indians were very troublesome, however, and, until 1867, very little if any, but farming and prospect work was done, the Indians having completely cleared the camp about two years prior to this time.

"In 1868 Mr. Jacob Hess of Manti, with a party of prospectors, pushed up in the Bullionville cañon and among other prospects located the Bully Boy, which has produced ore in paying quantities more or less ever since. The early settlers located in the cañon west of the present site, thinking the protection of the hills would afford relief from the Indians. They soon after removed to the more open and milder climate down where they are now located, and Bullionville was abandoned. The relics of this camp are plainly seen to-day. In the early days' development, work was necessarily limited, on account of transportation facilities, which even to-day are so very limited that were the Rio Grande Western to push in, Marysvale would come out as one of the richest camps in the west. The mine owners are now only waiting the advent of this much-desired facility to spend millions to develop the latent properties.

"Among the well-known mining men and capitalists are R. C. Chambers, L. U. Colbath, G. F. Dalton, L. E. Hall, E. J. Yard, F. A. Powell (of the Rio Grande Western), W. C. Hall, R. Warnock, George M. Scott, A. M.' Musser, Elias Morris, Henry W. Lawrence, Gilmore & Salsbury, Marshal Parsons, Ex-Marshal Dyer. Mrs. Senator Hearst, of San Francisco, besides a great many local capitalists who are not so well known as those above mentioned.

"To commence with, we will briefly note the mines which have been patented, which, of course, we will remind our readers, requires an outlay of $500 to secure this, and were there any question in the minds of the patentees as to the quality of the investment, we may be sure they would not plank down this much cash in an idle venture.

"The Bully Boy and Webster is the first location in the camp, and on it has been done more development work than any of the later discoveries. This property is owned by the Webster Milling and Mining Company, Messrs. R. C. Chambers, L. U. Colbath, L. Hale, Thomas Ferguson, all of Salt Lake, except the last named gentleman, who has full charge at the vale. They are located about six miles up the cañon, in the Ohio district. The mill works of this company contain a 10-stamp mill, but heretofore they have only run about 60 per cent of this capacity, but preparations are at present being made to increase this capacity to 18 tons per day. They will put in new concentrating machinery and will run to their full extent. The contracts are out for completion of the works by April 1, 1892, when work will be in good shape for operations on a large scale. The ore shows a regular milling assay of $60 in gold and silver, and there is considerable ore on the dump. All operations, except drifting and development work, are stopped until April 1st. They have a 105-foot fall for the mill power, with a turbine power of 80-horse. During the past season they have been running about eight tons per day in the mill and shipped about one ton of ore per month and two tons of concentrates. They have employed from eighteen to twenty men at the mines and seven to ten in the mill, running day and night. They have paid in wages about $3,000 per month. After April 1st, they will increase this to $5,000 or $6,000, and have an output of fully double that of 1891. This summer they have run in a tunnel at the foot of the hill about seven hundred feet on the ledge, and at present it is showing a perpendicular depth of 500 feet, and when they reach the full length of this tunnel they will show a depth of over three thousand feet. They will then reach the extension of the Dalton property.

"The Dalton Company comes in next, as their operations this season have been on no small scale. They have just completed and placed in operation a Huntington centrifugal roller process milling plant at an outlay of $8,000. They have employed from twenty to thirty men since the latter part of August. This mill is located about three miles above the Webster and presents a very animated appearance when in full blast. The mines are located three miles further up the cañon, and consist of the Pearl and Hard Cash, and a group of prospects that are being worked and developed. From the Pearl and Hard Cash mines they have been obtaining, on an average, about six tons per day. Mr. Emmanuel Poulson has had charge of this department, conveying the ore in sacks on a pack of twenty-one mules. His bill amounts to about $600 a month. In some recent finds, a vein of three and a half feet shows the enormous figures of $1,600.40 and $1,500.70, to the ton, and a regular car lot samples from $200 to $270 to the ton, mostly in gold, although there is a trace of silver. The Dalton Company has expended this season in machinery and development work over $29,000, and in the spring will produce some of the most valuable ores ever mined in the territory.

"The Copper Belt is located on the north side of Bullion Cañon and is a patented claim owned by Thomas Ferguson and some others, who have leased it to Parsons, Vandercook and S. F. Mount of Salt Lake. Mr. Mount has it in charge and has spent considerable money in development. He has sunk a shaft 325 feet, with good paying ore all the way down. The ore, as shipped in its crude state, is valued at from $500 to $600 per ton. The inaccessibility to the mine makes it very expensive transferring the ore to the shipping point. These gentlemen anticipate doing considerable work in the spring and will erect a concentrator.

"The Branch Lode, D. C. Tate, manager, has run in a drift about 50 feet and one about 36 feet, also a drift that follows the vein. They have taken out considerable ore, which remains on the dump. Mr. Tate is preparing to sort the ore for shipment. He will have some fifteen or twenty tons ready for shipment in January. In recent assays they found better ore. There are good, well-defined assay veins from 12 inches to 20 inches, assaying from $100 to $400, with about 20 per cent gold. This is owned by the Deseret Gold and Silver Mining Company, which is intending to drive a new tunnel at the foot of the hill to cross-cut the vein at its lowest point, which will make the mouth of the tunnel close to the road. Besides the original vein, there are three others, ranging from an average of about $40 to the ton, and will be mined as developments are made. This is the property owned by Messrs. A. M. Musser, Elias Morris, a Mr. Thompson and other prominent Salt Lakers.

"Uncle Sam is a patented mine owned by Mr. Bevans, of Salt Lake, with others. It shows a large vein of high grade ore. It is situated in the Mount Baldy district, about ten miles southeast of Marysvale. There are some valuable iron claims taken and patented in this district; they were originally located for fluxing purposes, and will eventually be valuable property.

"The Pluto is located in the Mount Baldy district and shows very high grade ore, but it is of very difficult access. Thomas Ferguson, with some others, are the owners.

"The Mountain Chief is a contact vein averaging about three feet, running about $500 to the ton. This comes in the Mount Baldy district, up the Cottonwood cañon about ten miles south of Marysvale. It is owned by King & Son and Mr. Ware –poor men awaiting the wheel of capital to turn their way.

"The Best Out is another valuable property showing low grade ore in plenty, and easy of access. It is owned by R. Warnock and others. They have a large consignment on the dump ready for railroad connection.

"The Clyde Group embraces a number of prospects and mines with assays running high in gold and silver, some of which show $50 in silver, $300 in gold and about 15 per cent in copper. The owners have done considerable work—about 400 feet of shaft and fully 1,160 feet of tunnel and drifting. These properties are owned by W. C. Hall, of Salt Lake, and Thomas Ferguson, and they have recently closed negotiations with an English syndicate, who are required to expend in development a working capital of $75,000 under the supervision of the owners. This will create a considerable boom early in the spring.

"The Crystal is another patented location in the Mount Baldy district. Over $25,000 has been expended in development and about ten thousand tons of ore is on the dump. This is owned by Mayor George M. Scott, of Salt Lake, and shows encouragement in every respect It is understood a large force of men will be put on in the spring and extensive operations extended.

"The Grand View is owned by Dewitt, Robbins & Hammil, and shows a sample assay of $60 to the ton. It is located in the Ohio district.

"The Star Group is owned by R. Warnock, of Salt Lake; $20,000 has been expended in development work, and sample assays show a valuation of from $20 to $500. It is located in the Ohio district, about a half mile west of the Bully Boy and Webster mines.

"The Cascade is located about the same as the Star Group and shows an assay of from 20 to 150 silver and 22 per cent lead. It is owned by Miles Durkee and Reuben DeWitt. The vein is about four feet and is of easy access. A force of men will be put on early in the spring, if capital can be secured.

"THE MOUNTAIN QUEEN GROUP.—This valuable group is owned by the same parties as the Cascade, and is another one requiring working capital. The owners have done considerable prospect work and show an assay of from $60 to $460. It is located near the Dalton in the Ohio district. They have a shaft, cross-cuts and tunnels, and mining experts say this is one of the best properties in the camp.

"The Kearsarge, located over 2,000 feet west of the Bully Boy and Webster, runs about $62 in gold. This is owned by E. J. Yard and F. A. Powell, of the Rio Grande Western, and W. H. Cowans. The development work shows a shaft seventeen feet, a tunnel about thirty feet and a vein averaging about one foot. About ten tons of good ore are on the dump.

"Lost Lead is another poor man's property, and one of the most valuable in the camp. It is owned by L. Washburn and Frank Wright.

They have run a fifty-foot tunnel and sunk a shaft about one hundred feet, with fully 60 feet of drifting, showing a body of free milling ore running from $40 to $65 in gold and silver. They have expended about $3,000, two-thirds of which has been used within seven or eight months. This is located near the Bully Boy and Webster, in the Ohio district.

"The Pistol Pocket is a find made by Frank King, and our report, which appeared in the daily of October, met the eyes of a capitalist, who is putting up the capital to push developments this winter. It is in the Ohio district, of quite difficult access. The general outlook is fully 100 per cent better than ever before, and from the present indications when spring opens up, Marysvale will be the liveliest camp in the territory in proportion to the population. This, of course, will be largely augmented when the Rio Grande Western gets in, which is expected by June 1st, 1892.

"The Antelope, owned by Febyre, McCarty, John Jacobs and Mrs. Stark, is situated in the Henry district in Sevier County, about five miles north of Marysvale. It has a gold and silver bearing ledge. The pay streak is about three feet, assays about $27.60 on an average, and carries 25 per cent lead, 25 in gold, and 70 ounces in silver. They have a shaft about 80 feet deep and some drifting, amounting to about 4,000 feet. This property is of easy access, and an old miner says he will place 5,000 tons on the dump at $5 per ton. The dump is within about a mile from the surveyed road of the Rio Grande Western, and a cañon road leads to the mine.

"The Sevier Gold Mining Company is controlled by Salt Lake parties. They have several valuable claims, and have shipped some ore. A ten-stamp mill has been erected, and shipments of gold bullion may soon be looked for.

"Reed & Benson is an old property that has shipped heavily in the past, and will soon resume its former prestige. A complete electric plant is now being put in under the management of H. C. Goodspeed.

"A force of men has been busy on the Silver King during the greater part of the year, but no shipments were reported."

WASHINGTON COUNTY.—From the same source we extract the following interesting chapter:

"Few people realize the vast amount of mineral wealth contained in the mountains of Washington County, now lying comparatively dormant for lack of railroad facilities. A list of the advantages that would accrue to the people of Southern Utah by railroad connection, either with the North or West, would fill a volume. Among the many benefits to be derived by such connection, a few may be mentioned. The advent of the iron horse into 'Dixie' in the South means renewed energies on the part of the people; the development of unlimited mineral resources hitherto only partly worked; the opening up of new agricultural lands second to none in the world for productiveness, and only awaiting home-seekers and markets for the yield; an increase in the shipment of cattle, horses, sheep and hogs, of which the southern tier of counties are the greatest producers in the West; the erection of great sanitariums at the many mineral springs of Washington County, the waters of which possess as high medicinal qualities as do any in the world, those of Carlsbad and Arkansas not excepted; an increase in the acreage devoted to vineyards, orchards, of early and semi-tropical fruits, and the production of cotton; the establishment of canneries and various other manufacturing plants; and last, but not least, would bring the people into close communication with the outside world, thus opening up new territory for business transactions and commercial relations.

"Nature seems to have intended the country below the rim of the Great Basin as her mineral treasure house, and deposited therein unlimited quantities of useful and precious metals, which, as the territory north, east and west is fast reaching that stage when investors in search of good investments for their wealth must turn to pastures new, is beginning to attract the attention of mining men the world over. The country in this district is of volcanic origin, and the earth in its tremendous convulsions seems to have turned inside out, and left great ledges and bodies of ore exposed to the view of even the most careless observer. The mountains of Washington County are filled with gold, silver, copper, lead, sulphur, iron, platinum, zinc, ochre, salt, alum, and mineral wax in unestimated quantities. Almost every mineral known in mineralogy has been found there, but the lack of cheap transportation makes it almost impossible to develop the prospects to their full extent, and valuable properties that would yield millions if they could be properly worked, are now lying idle.

"Within the last year mining in the South has received an impetus never before experienced since the great silver camp, Silver Reef, practically ceased operations. Chiefly through the efforts and operations of Woolley, Lund & Judd, Washington County has again become a producer to no mean extent, and now has in full operation one of the richest copper mines in the country—the Apex, one of the Dixie group—situated about eighteen miles southwest of St. George, the county seat and metropolis of Washington County. The group is owned by Woolley, Lund & Judd, and is incorporated under the title of the Dixie Mining and Smelting Company, with John Pymm, president; Robert C. Lund, vice-president; Seth A. Pymm, secretary; James Andrus, Thomas Judd, H. Pickett and Brigham Jarvis, directors. The company is the owner of thirteen adjoining claims.

"The first discovery was made several years ago by St. George parties, the outcropping or apex being nearly to the top of the mountain. A shaft was sunk a considerable distance until it became evident that the vein went down at an angle of nearly 45 degrees, and could be reached from below. As a large amount of ore of good paying quality had been taken out, a small smelter was erected, and made several short runs with good success. The mine then lay idle until about a year ago, when Woolley, Lund & Judd took the management.

"For the purpose of more fully developing the claims, the company decided to go down the mountain about three hundred feet and run a tunnel into the hill in such a manner as to strike the vein where the ore body was more extensive. One hundred and fifty-five feet from the collar of the tunnel a large cave was discovered. A man was let down on a windlass 300 feet, and made a full exploration of the cavern, but could find no connection whatever with the ledge or ore body. A strong current of air came from below, thus insuring the best of ventilation for the mine and also good drainage. After about fifty feet the ore body was encountered, the grade of rock being good. In the hope of striking the ledge of galena, which cropped out near the copper at the Apex, the tunnel was continued until it had reached a depth of 400 feet. The miners were then withdrawn from that portion of the mine and put to work on the ledge proper, which is over one hundred feet thick, composed of low grade carbonate lead ore, soft iron and decomposed lime, with occasional bunches of rich galena. The high grade copper ore, which assays as high as 85 per cent, is found in almost inexhaustible quantities near the hanging wall of the mine, a ledge of dolomite. The country rock on all the claims is a hard, dark-colored limestone. At present five large ore slopes are in sight, and turn which way you will, high grade ore is encountered. Since work was begun on the tunnel in November, 1890, the company has worked an average of six men, sometimes running as high as twelve.

"The high grade ore is all shipped east to New York, but in order to reduce that of lower grade, the company erected a twenty-ton water-jacket smelter at St. George, on the creek which furnished the people with water. The blower and crusher are run by water power, a good head being obtained by building a new canal at a considerable expense. The mine furnishes all the flux necessary to reduce the ores, which smelt very freely.

"The smelter started up October 26th and run sixteen and one-half days without an accident, reducing 350 tons of second-class ore and producing 90.63 tons of bullion. The run proved so satisfactory, both in workings and financial results, that the company is preparing to continue work, and has about 300 tons of ore at the smelter. As soon as enough charcoal and material are on hand, another and lengthier run will be made. The copper sells readily, being of very superior quality, and classed with the lake product.

"The output since the active work was prosecuted in June, until December 1st, is as follows: Six cars of first-class ore (199,770 pounds), average fire assay 48.23 per cent, sold in New York for $8,407.61; one car second-class ore (23,130 pounds), average fire assay 35.85 per cent, sold in New York for $674.04; 173,710 pounds bullion, fire assay 86.7 per cent, sold in New York for $12,951.81; 7,558 pounds matte, fire assay 65.7 per cent, sold in New York for $402.19; making the total value of the product in round numbers, $22,435.65.

"Situated about one mile north of the Dixie group, and in the same chain of mountains, is the Mammoth mine, in which was struck a large cave, filled with high grade carbonate lead ore, carrying from twelve to twenty-five ounces of silver. The Mammoth is controlled by Woolley, Lund & Judd, and is one of the finest prospects in Utah. If situated on the railroad, the mine would furnish freight enough to keep one road busy, and from all indications the supply of mineral is inexhaustible. The walls and floor of the cave are covered with the rich ore which can be shoveled up in great quantities. It is only a question of time when the Mammoth will become one of the Territory's biggest producers.

"The St. George Mining Company, a corporation formed in Omaha, whose stockholders are among the wealthiest in the country, owns several fine prospects near the Dixie and Mammoth claims, but up to date has done but little work. The avowed intention of the company is to erect large reducing works and use its influence toward the building of a railroad into the district.

"Silver Reef, the great Southern camp, one of the greatest agents in the building up of Southern Utah, has been quiet during the past year, and has little prospect of improving until silver becomes more valuable. All the work done there this season has been by chloriders. The old Barbee mill made two short runs on custom ores. The Christy also made two runs on ores taken from the Christy Company's mines. The mines of the Reef are among the richest in the country, and, beyond a doubt, the camp will in the near future resume its old-time activity. There are millions yet lying in the white sandstone reefs which, in time gone by, have proven such bonanzas to the owners of good prospects. The producing mines of Silver Reef are the Old Buckeye, Last Chance, Thompson and McNally, Barbee and Walker, Tescumseh, California, Stormy King and Neutral."

WEBER COUNTY: LA PLATA.—To the northeast of Ogden about twenty-two miles is the new mining camp of La Plata, which was ushered into existence about sixteen months ago.

The mineral was discovered by a sheep-herder. While riding over the hills in that vicinity, the horse which he had broke a chunk of ore off of a solid piece of galena which was exposed to the surface. The sheep-herder little realized then that the very spot where he was standing would furnish a field for the prospector and the miner. He put the ore in his pocket, and when his employer was next seen he showed it to him. He took the piece, and the herder showed him where he had picked it up. This induced the sheep man to begin prospecting, and with little exertion great bodies of ore were uncovered near the surface. The news soon spread abroad, and, within a few weeks, the surrounding hills were alive with prospectors and miners. A town site was selected in what is known as Bear Gulch, and named La Plata, from which the camp derived its name. It now has a population of over 125 people. Each day brought the tidings of new discoveries, and, within the period of a few weeks, several hundred locations were made.

The La Plata mining regions, so it is said by old miners, very much resemble Leadville, but so far its productiveness is something of a disappointment.

The hills are not high nor steep. A dense growth of timber covers them, and great springs of water come flowing down their sides. Above the town of La Plata, a large spring of water comes bubbling from the ground, sufficient to supply a city of 10,000 souls.

The country surrounding Park City is very much the same as La Plata, only the absence of timber.

OTHER MINES.—Many other old and new discoveries near Ogden are being worked, but not as vigorously as their owners intend to be the case hereafter.

SAN PETE COUNTY.—There have been several locations made of gold, silver and lead-producing mines in San Pete valley, some of which have been worked considerably. The one showing the greatest development, and the only incorporated property is the Alexander, which modestly puts its capital stock at $45,000, divided into 9,000 shares of $5.00 each. Over $1,500 worth of work has been done, and a tunnel is now being pushed to cut the ledge at a depth of about 150 feet. Ore has been obtained running as high as 913 ounces in silver. The officers and directors are: F. R. Kenner, president; A. H. Cannon, secretary and treasurer; S. A. Kenner, attorney; Frank J. Cannon, Beauregard Kenner and F. J. Nelson. Work has not been prosecuted vigorously of late, but is expected to be shortly.

San Pete has great coal measures, and has been a producer of that article in considerable quantity for years.

OVER THE LINES: PIOCHE.—Pioche is reached to best advantage from the terminus of the Union Pacific, at Milford, and so is Osceola, the former south, the latter west, of Milford, in Nevada. At Pioche, work on the mines during the past year was more in the line of development, of opening ore bodies, and in other ways preparing for steady shipments, than in stoping or making a showing in the way of output. The Pioche and the Yuba companies were consolidated as the Pioche Consolidated. They own several large groups of mines, comprising most all the old producers of note—the Raymond & Ely, Meadow Valley, Mazeppa, Newark, American Flag, Hillside, and Day—together with a number of newly discovered mines—Half Moon, Mendha, and Onondaga. Ores, sulphides and bullion shipped by the company since it began operations amounted to over $104,071.64. It is found upon trial that the ores are remarkably well adapted to smelting. There is a combination of silicious, iron, lead and lime gangue ores which form a self-fluxing smelting mixture; such a complete variety as had not been found before in quantity within the same radius (ten miles) in the United States. Mr. August Werner has charge of the smelting, and he has succeeded in smelting in one stack, with the high prices of labor, fuel, etc., incident to being 110 miles from a railroad, for only $8.16 per ton, running an average of over 50 tons per day through, with a 10 per cent lead charge, and making 200 ounces of bullion and the cleanest of slag. The slag is something new in metallurgy. To make sure if the reported daily slag samples were properly taken and that such uncommon results were correct, several large samples have been taken from all parts of the slag dump, "cut down" and assayed by different assayers, and they have more than confirmed the results, the three general samples having assayed as follows:

	Ozs. Lead.	Ozs. Silver.
No. 1	1	Trace.
No. 2	0.8	Trace.
No. 3	None.	None.

As soon as the railroad reaches Pioche, the Salt Lake smelting interests will be greatly benefited by the superior fluxing ores of that region. The lime ores especially are an important feature, as that is the character of flux in ores now lacking in Utah. A regular supply of this ore would save the quarrying and smelting of barren limestone in Salt Lake, and thereby cheapen smelting and help keep ores there for treatment that are at present being shipped east. This lime ore exists at Pioche in apparently inexhaustible quantities. Prof. Geo. W. Maynard, who has examined the Pioche mines, estimates the reserves of the lime fluxing ore in one mine (the Day) at 482,000 tons. This means a shipment of 200 tons per day for over six years, in sight. The ore is a mineralized limestone, being about two-thirds carbonate of lime, with the remaining one-third made up mostly of oxides of iron and manganese. It contains only 3 per cent silica, and carries about 3 per cent lead and twenty ounces silver per ton.

Prof. Maynard also carefully measured the shipping ore in the other mines at Pioche, and found 72,624 tons, making in all over 500,000 tons "in sight," which indicates that, although Pioche has not made much noise, the district has been earnestly at work the past year, and it now warns the mining world that it intends to beat its old record for good ore and plenty of it.

OSCEOLA.—The Osceola Gravel Company, early in 1890, completed a ditch 18¾ miles long, with a capacity of 2,500 miner's inches, or 40,000,000 gallons per twenty-four hours. The ditch has a fall of 16 feet per mile. The old ditch, brought from the opposite side of Wheeler's Peak, is about 17 miles long, and has a capacity of about 2,000 inches. The two ditches delivering water in the same gulch, furnish a great supply. Washing began in March and continued until December 10th, when the cold weather caused the monitors to be shut down. In operation two monitors are run at a time, there being two nine-inch and one seven-inch. Fifteen men are employed during the season in the mine. This gravel bar has been prospected over hundreds of acres, and estimated to average 17 cents gold per cubic yard, but in operating it has run as high as 27. About 200,000 cubic yards of gravel were washed out during the season, of which only a part of the gold was taken, because, this being the first season, much gold was left back at the base of the banks on bed-rock, and which will be caught in the sluiceway next season. In starting the gravel was thin, but going upward gained in depth until the face of the bank is now 92 feet high. Water is sent against this bank under a pressure of the monitors of 225 feet. The bed-rock flume, or sluiceway, is four feet wide and four feet deep, and runs full most of the time. This is about 300 feet long. The old ditch supplies power for operating a 2,000-candle-power electric dynamo to furnish light for the workmen at night, and then this water goes back into the ditch to help wash out the gold. The company cleaned up a handsome sum, but declined to name it to the public.

METAL OUTPUT FOR 1891:—

1,836,060 lbs. Copper @ 5½ cents per lb..$	100,983 30
6,170,000 lbs. Refined Lead @ 4 cents per lb....................................	246,800 00
80,356,528 lbs. Unrefined Lead @ $60 per ton................................	2,410,695 00
8,915,223 ozs. Fine Silver @ 98¼ cents per oz................	8,759,206 59
36,160 ozs. Fine Gold @ $20 per oz...	723,200 00
	$12,240,885 73

Computing the gold and silver at their mint value and the other metals at their value at seaboard, it would increase the value of the product to $16,198,066.81.

Six thousand tons of Iron were shipped from Tintic for fluxing purposes.

COST OF MINING AND MILLING.—This varies greatly with circumstances. At the Ontario it is something less than $30 per ton; at the Daly it is given at $26; at the Horn Silver in 1883-84 it was about $26. These figures include all cost for the year, maintenance of plant, dead work, incidental expenses, but not, of course, original cost of plant and opening of the mine. At Silver Reef, cost of mining and milling is $13 to $15. The mass of Utah low-grade ores requires concentration, but this costs only about $1 per ton. For every ton of concentrates, however, 3 tons of ores must be mined and carried to the concentrator. Bingham and Stockton and Ophir are the low-grade districts; Park City, also, in part. Part of the low-grade ores have to be roasted, the lumps in out-of-door heaps, the fine in reverberatory or revolving roasters. The ores milled at the Ontario and Daly have to be roasted and chloridized, while the dry ores of Tintic must pay heavy working charges. The figures given are the cost figures of mines varying widely as to location, natural conditions—as dimensions of vein or ore bodies, water, distance from market, etc.—grade and nature of ores, appliances and processes of reduction. But doubtless $30 per ton amply covers cost of extraction and reduction of all Utah ores.

SAMPLING AND SMELTING.—There are thirteen sampling mills in Utah—one at the Horn Silver mine, one at Milford, one at Tintic, five at Sandy and vicinity, three at Park City, and one in Salt Lake. Together they sampled in 1890, about 150,000 tons of ore. Ordinarily, only the fifth or tenth sack of a lot of ore is sampled, and the cost is $1 per ton for the whole of it. Where the whole is sampled, the charge is $4 per ton. The sampler crushes the ore to the size of peas, thoroughly mixes, and sends sealed packages to the assayer, upon whose certificates it is bought and sold.

In the Jordan Valley, six to twelve miles south of Salt Lake City, on the railroads, are the Utah smelters, four or five different concerns, comprising about a dozen stacks. Those in blast at present are, the Germania, three stacks, three revolving roasters, and one large reverberatory; the Hanauer, 4 stacks and 5 roasters; the Mingo, 4 stacks and 5 reverberatories; the three plants valued at $500,000. Together they keep 7 or 8 stacks pretty steadily in blast, and employ about 350 men at an average wage of $65 per month.

About one-fourth of the Utah ores were shipped out of the Territory for reduction. There is a good opening at Salt Lake for a great smelting works.

STOCK EXCHANGE.—A Stock Exchange was organized in Salt Lake City in 1889, and held its first session June 6th. Thence to the end of December, sales were 2,375,275 shares, total consideration for, or value of which, was $647,837.56. Sales of silver certificates in same time were 896,000 ounces; consideration, $940,800. This is not a very great business, but it was believed to be of advantage in advertising the mines, securing money for their development, and a great convenience in fixing the value and providing a handy market for mining shares, locally considered. It is silent at present, waiting, perhaps, for a renewed activity in silver mining.

MINING IN GENERAL.

COAL.—Utah contains a great variety of minerals besides those involved in silver-mining, to-wit: Silver, gold, lead and copper. Coal occurs on both fronts of the Wasatch, and of the High Plateaus, almost the entire length of the Territory. The coal measures underlie an area of many thousand square miles; probably 2,000 that are available. At all events, there is enough to meet any possible demand for generations. We should be mining four times as much as we are, but the Union Pacific largely supplies Utah from Wyoming.

The Union Pacific owns coal mines in Pleasant Valley (Scofield), from which the past year they mined of commercial coal, 88,000 tons, and probably as much more for their own use, making in all about 200,000 tons.

The Home Coal Company raised and sold in 1891, from their own mines on the Weber, near Coalville, 65,138 tons, and the Chalk Creek Company, from mines also near Coalville, which raised and sold 1,200 tons in 1890, largely increased its business; Salt Lake City consumed 88,400 tons during the eleven months of the present year (1892) it has already consumed greatly in excess of that sum.

SUNDRY MINERALS.—There are deposits of brimstone near the mouth of Cove Creek, about 30 miles east of Black Rock Station, on the Union Pacific Railway. This deposit is supposed to be practically inexhaustible. There is a deposit at Hilliard, another about 12 miles from Frisco, and still others.

Ninety miles from Juab Station, on the Union Pacific Railway, up the Sevier river, at a place called Antimony, deposits of antimony ores were worked. Such as could be reduced without concentrating were exhausted; in the construction of concentrating works, costly mistakes were made; the company's money gave out, and work ceased. The antimony turned out was of extraordinary purity, and, with railway facilities, operations may be resumed. There are said to be available deposits of antimony ores in other parts of the Territory, especially in Boxelder Cañon.

Quicksilver ores are found at Marysvale, and also at Lewiston. Bismuth occurs in Beaver County, east of Milford, and also in spots in some of the mines of Tintic. Copper ores are found at Bingham, at Tintic, in North Star, near Frisco, on the Cottonwoods, in Lucin District, Boxelder County, at Deep Creek, all over the Territory, in fact.

Iron ores are found about Ogden, in Morgan, Boxelder, Cache, Salt Lake,

San Pete, Tooele, Juab and Iron Counties. The iron mines above Willard furnished ores for fluxing purposes in early times. For many years 6,000 to 12,000 tons have been yearly drawn from Tintic by the smelters for fluxing silicious ores. The deposits in Iron County, about 300 miles south of Salt Lake City, are amongst the noted deposits of the world; at least they are so considered by authorities on the subject.

They are scattered about in a belt two miles wide by sixteen miles long, in number about 50, and with very little work done on them show about 3,000 tons of ore in sight. Twenty-three samples taken by an iron expert, known to the writer, showed upon analysis an average of 65.98 per cent metallic iron, .042 per cent phosphorus, no trace of titanic acid, practically no copper, and a residue, mostly silica, of 3.6 per cent. In some of the samples there was a little carbonate of lime and also manganese. Following are the best samples so far as absence of phosphorus is concerned:—

ORE IN SIGHT.	Met. Iron.	Phosphorus.	Residue.
100 x 35 feet...............................	67.2	.100	2.2
85,714 tons........	63.8	.016	4.5
20,857 tons....	68.8	.041	1.9
8,571 tons	69.1	.044	1.1
41,428 tons	62.3	.005	6.8
1,535,569 tons	68.9	.055	2.8
31,546 tons......................	69.5	.034	2.2
71,471 tons	69.0	.011	2.5
34,286 tons...................	67.2	.049	4.3

These figures need no comment, and are under rather than over the score.

Deposits of rock salt, some of them quite pure, are found near Nephi, on Salt Creek, and also near Salina, and in other localities. It is useful in its crude state for feeding stock, for chloridising silver ores, and it may be refined and put to all kinds of use. Great Salt Lake is an inexhaustible storehouse of common salt, and, the chemists say, of a variety of sulphates, borates and bromides, from which may be manufactured salt cake, epsom and glauber salts, soda ash, bi-carbinate of soda, caustic soda and sal-soda.

HYDRO-CARBONS.—Curious and valuable hydro-carbons are found in the Uintah-White Basin, and about the Pleasant Valley Divide. A company with headquarters at P. V. Junction is mining for ozokerite (paraffine). So far the mineral has not been found in large quantity. It occurs in seams, bunches and stringers, where the material has been caught when in a volatile state and held till it condensed into a solid.

Gilsonite or Uintahite occurs in the bad lands of the lower Duchesne and the lower White, in veins or lodes striking straight through the sandstone formation, standing vertically, thirty inches to twenty feet, thick, clean, black, and, when first broken, lustrous as jet. It is 99½ per cent asphalt, with the oils dried out. Most of the known veins are on the

Indian Reservations, but one of them has been set off by Act of Congress, and is owned and wrought by a St. Louis company. It is used, as yet, mainly for varnishes, but it is expected that more extended use will be found for it.

On the Green River and eastward, asphalt, and oil with an asphalt base, exude in places and form deposits said to be not unlike the asphalt lake at Trinidad. Prof. Newberry is of the opinion that this is a petroleum region. Petroleum, he maintains, is derived from the spontaneous distillation of hydro-carbons, and as the Colorado group east of the Wasatch consists of bitumious shales 1,500 to 2,500 feet thick, gas and oil springs are to be expected. The gilsonite and the ozokerite Prof. Newberry refers to this distillation. It is probable, he says, that these residual products of the liquid hydro-carbons evolved from the shales, as well as petroleum, will become important items of export from this region.

A kindred substance to these, which Prof. Blake, of New Haven, names "Wurtzilite," has been found about the divide between the Strawberry and the Price, close up to the Wasatch Range. Before it could be located, and its extent ascertained, it was discovered that it was mainly on the Indian Reservation, and prospectors were warned away. It is of no use to the Indians, neither is the extremely high and broken country where it is found. The latter should be restored to the public domain, so that the arts may have the benefit of this material if it can be used. Its mode of occurrence is somewhat like that of ozokerite, but it is more plentiful.

BUILDING STONE.—Structural, fertilizing, and abrasive materials of every variety, and adopted to all uses, are found all over Utah, and generally convenient to the valleys where the people live. A number of stone quarries have been opened the past year. The Diamond, Kyune & Castle Stone Company worked quarries of brown sandstone at Diamond, and of gray sandstone at Kyune, and at Castle Gate, all on the line of the Rio Grande Western Railway. Their shipments the past year were enormous. Cars go to Seattle (Wash.), Logan, Milford, Nephi, Ogden, Salt Lake City and numerous other points. Cubes of these stones, tested at the Illinois State University, cracked—the gray Kynne stone under a ten minutes pressure of 16,000 per square inch, and broke under the same of 20,800 pounds; the brown Diamond stone cracked under a pressure of 30,000 pounds, and broke under a pressure of 34,550 pounds. Twelve cubic feet of these two kinds of stone weigh one ton. Excellent foundation and dimension stone is brought into Salt Lake City from the adjoining cañons and from Parley's Park. All the larger towns of the Territory find the best of building stone, and, it may be added, the best of clay, except Kaolin, at their doors, so to speak. Lithographic stone of good quality; marbles; gypsum; slate; the materials for the manufacture of glass, and of Portland cement; rock rich in asphalt; limestone for building and for fluxing ores—these materials are found in many places in the Territory. A gypsum mill near Nephi is sending plaster to various points on the Pacific Coast. Salt is made and gathered on the shelving shores of Great Salt Lake, and supplies the chloridizing silver mills of Utah, Montana, Idaho, and part of Nevada.

NATURAL GAS.

One of the greatest recent features added to the commercial prosperity and general attractiveness is the discovery and development of natural gas in immense quantity at a point practically within the borders of Salt Lake City. It has been known for several years that the city and its surroundings were underlaid with veins or deposits of that needful article, and in many cases it had been developed and applied, but only in a small way, and with primitive apparatus; this, however, sufficed to keep the interest alive and on the increase, and finally, on February 24, 1891, the Natural Gas Company was incorporated, with a capital of $5,000,000—five hundred thousand shares at $10 per share. Mr. James F. Woodman, the well-known mining man, is president of the company, and the other officers, several of whom are eastern men—though the controlling interest lies in Salt Lake—are as follows: President, James F. Woodman; first vice-president, W. A. Nelden; second vice-president, Schuyler C. Constant; treasurer, H. L. Driver; secretary, I. T. Stringer; Joseph J. Rogers, Wendell Benson, P. L. Schmidt. Several other companies have since been organized.

Their leases being secured, the company went to work sinking where there had already been best indications of gas near Farmington, a few miles northwest. As labor progressed, it became evident that the whole surface of the earth thereabouts was impregnated with natural gas. About the beginning of December, 1891, the main well, known as "No. 1 Gusher," burst forth with a volume of gas and smoke that sent boulders, shale and dirt fully three hundred feet into the air. Ever since then the up-pouring of gas has continued, and the problem has been to get it under control; this has now been accomplished, and recently many hundreds of people have visited the wells, and many experts inspected it. On December 17th, the mayor and city council, with several hundred citizens, paid the wells a visit, and on all hands expressions of wonderment and confidence in the permanency of the flow were heard.

The company now have pipe line franchises with Ogden, Salt Lake, Provo, Centerville, Farmington, Kaysville, Bountiful, Sandy, and many other small towns. The company have reserved the right to purchase reservoir lands on every farm leased. The rights of way are over a level and a valuable fruit, farming and manufacturing territory. Millions of tons of raw material lie contiguous to the wells and pipe lines.

Three railroads pass through this company's property, Union Pacific, Rio Grande Western, and the Hot Springs line, two of which are transcontinental lines, while their branches reach every mining camp of prominence in Utah, and are connected by one grand system in Idaho, Colorado, Wyoming, Montana, Oregon, Washington, Nevada and California, while work is being done to connect New Mexico, old Mexico and Arizona, all direct with Salt Lake City, Ogden and the oil and gas fields owned and controlled by the American Natural Gas Company.

The company have two more well-boring outfits on the way from the east, and they will now rush the sinking of wells. These wells will be sunk in a line towards this city, at a distance of a half mile apart. With such a vast field of gas and an outfit so well equipped for sinking wells, they propose to sink a dozen or more wells as soon as possible and get them capped, while they will put in a pipe line to Salt Lake City at an early date. The plan is to have enough wells to let part of them remain resting while the others are in service.

The gusher is down 600 feet, and flows from a 6-inch pipe. A test has been made of its pressure, and 350 to 400 rock pressure recorded. Some idea of its force and strength, however, can be obtained from the account of the big fire which took place there on Dec. 12, 1891. The following is taken from an account published next day in the Salt Lake *Herald:*

"The great six-inch gas well reached down to a depth of 550 feet yesterday morning, penetrating to what is understood as the Trenton sandstone formation, where gas in large quantities has been found in other parts of the country. According to statements of responsible parties who were at Lake Shore yesterday, the pressure increased to 150 pounds to the square inch, and when the big valve was opened, the gas rushed out with a roar that was heard for over a mile distant, and people came hurrying from all around to ascertain the occasion of the uproar. Unfortunately in the course of the work the big well got to leaking, and immense quantities of gas siped up through the soil for a distance around. Through some cause, very likely a lighted cigarette, or live coal from the furnace, the escaping gas caught fire, and—whiff! the whole country seemed to be in a blaze. Manager Smith, of Kansas City, and one of his assistants were caught within the circle of fire and they had to do some tall leaping and running to escape, and escape they barely did with their lives. As it was, they lost their eyebrows and part of their hair, besides being badly scorched on the face and hands. The gas came out of the ground with considerable force and the flames shot up to a great height. They presently caught the plant buildings in which were the engine, boiler and other machinery used on the place, and in a jiffy $500 worth of property was in ruins. The flames appeared to increase rather than diminish, and the manager and his men set resolutely to work heaping dirt on the fiery furnace. But that had no perceptible effect. So men were called in from around the neighborhood to help. Even that was insufficient, and messengers were sent further out into the country for assistance. In a few hours there was a big force of men on hand throwing dirt on the belching flames. Messrs. Larned, Irvine, Newell, Benson, Pettengill and other citizens who had gone up from this city on the 3:30 p. m. train could see the flames shooting heavenward when the cars were two miles from the spot, and the heat was something fearful when they got into the grounds. When they returned to this city early in the evening the fire had not been gotten under control. It was not expected that the flames would be extinguished before this evening, and it may be necessary to use carbonic acid producing chemicals in order to extinguish the conflagration. The fiery spectacle made a great sight for the passengers on the passing trains."

There is a third well on which sinking is now being done, with good indications.

What the future of natural gas is, in Salt Lake, it is impossible to say at this writing, but from what has been said it is not too much to expect that this interest adds one more to the many vast resources that are destined to make of Salt Lake the great overshadowing city of the inland Western America.

MANUFACTURES.

IN THE TERRITORY.—From the report of the Territorial Statistician, published in October, 1891, a very comprehensive idea is had in regard to manufactures in the Territory. He states that there are 3,974 operatives employed in the different concerns in Utah, and that the annual product turned out amounts to the value of $5,836,103. In the Territory, up to the closing of his report, the Statistician estimates that there are in operation at the present time 54 flouring mills, 42 saw mills, 11 planing mills, 10 foundries, 6 woolen mills. There are also carding mills, lath and shingle mills, ore samplers, concentrating and chloridizing mills, roasting and smelting furnaces, paper mills, clothing, boot and shoe, hat, glove, hosiery, silk, broom, brush, sash, door, blind, cracker and vinegar factories; iron, glass, soap, furniture, chemicals, and cooperage works; potteries, tanneries, boiler works, and many other kinds of business classified as manufacturing in the report.

IN THE CITY OF SALT LAKE.—There are in successful operation boot and shoe, knitting and overall factories, woolen and paper mills, tanneries, confectioneries, fence and mattress factories, cracker factories, show-case makers, brick makers, aerated water works, roller grist mills, cigar factories, vinegar factories, soap making, salt refining, chemical works, glass works, wood working, printing, book binding, brewing, etc., which give employment to about 500 operatives, and the amount of money invested in these concerns is $3,107,000, and the wages paid amounts to about $200,541. The merchantable products amount•to over $4,500,000. This is a very excellent showing, and the possibilities of the Territory have hardly been touched.

The Salt Lake Chamber of Commerce made a report on manufactures some four years ago, in which occurs the following:

" We find that all ventures in this city for the utilization of our surplus capital and natural resources have been successful and paid gratifying dividends, save where gross carelessness or incompetent management were displayed or where want of necessary capital was manifest.

" In many of these industries we find what would otherwise be thriving, labor-making, and money-saving concerns languishing for want of a little capital with which to improve their plants, advertise their wares, and place their products on a ready market. In other directions, notably in the manufacture of sugar, window glass, leather, paper, putty, cement, candles, brushes, paints, white lead, sheet lead and lead pipes, agricultural implements, spirits, medicinal preparations, earthen sewer pipes, canned goods, pickles and sauces, pails, tubs, kegs, barrels and step-ladders, wagons and carriages, stoves, baskets, demijohns, clothing, hats, etc., and in the successful operation of lithographing establishments, cigar factories, publishing houses, binderies, rolling mills, reduction works, manufacturing tin shops, wire working, and stone and marble sawing and carving, we find that capital can be so successfully employed in this city that it is a marvel to us that the opportunity has not been taken advantage of."

ADVANTAGES OF MANUFACTURING.—While it is true that the Territory has made a very good start in manufacturing, yet it still needs more manufacturing worse than any and all things else, if it would drain other towns and states of their money instead of being drained of its money by them. Agriculture and mining will not make a community rich. It is the working up of these raw products into fine materials, putting brains into them, that enriches a community. Manufacturing and commerce—transforming crude substances into articles of value and beauty, distributing and selling them—these are the indispensable requisites of wealth and prosperity. Transportation has passed into the hands of a class, and it is furnished as fast as it is needed, if not faster. Agriculture will take care of itself. Anybody can live by cultivating the land, and that is about all that anybody can do. Mining has its peculiar fascination. All the mines that will pay, and a great many that won't pay, are sure to be found and wrought without any urging. But manufacturing needs fostering. and within bounds it is proper that the manufacturer should be the favored man—so far as it is possible, favored of the community, favored by the Government. It is poor policy that sends a hide from Utah to Boston to be made into shoes and then returned, robbing Utah people of the work, and paying some one in Boston $5 to make the shoes, and some railroad owned in Boston for carrying the hide to Boston and carrying the shoes back. The money so paid away never returns.

FACILITIES: RAW MATERIALS.—It is true that all aricles which might be grown or made in Utah, and which are now purchased elsewhere and brought to Utah, have a protective (transportation) tariff of $10.00 a ton and upwards, in favor of their production in Utah. This tariff is against Utah in sending abroad what the Territory has to sell, which is an additional reason why advantage should be taken of it when it is in the Territory's favor.

What, then, has the Territory in the way of raw materials, and what about the other conditions of the problem? There is no Territory or State of the Union which possesses, or can produce, the raw materials of many important manufactures in greater abundance than Utah. For example, Utah has competition in local transportation; a climate that hardly interferes with out-of-door operations the year round; cheap fuel, coal and coke: good and cheap food produced at home; an ample supply of labor; much of it skilled labor, at fair wages. These are the prime conditions of the problem. Some of them are not altogether satisfactory, but they will tend to become so; that is to say, there will be more and cheaper transportation, a great deal more; there will be cheaper fuel; the tendency is that way.

IRON MAKING.—Now as to iron and steel, and all their secondary products, there are large deposits of rich and pure iron ores distributed down along the range from Cache County to Iron County. There is pure limestone in every hillside. What else is required save capital, skill, and a market? Utah can as well make iron and steel, and all the articles into which they are transformed, and send them to Chicago to market, if need be, as Chicago can make them and send them to Utah to market.

Alabama iron making (about $8.50 a ton) is the cheapest ever known. But Alabama pig iron could not be carried to Utah without enhancing its cost by $20.00; and this with the original cost would make the value of Alabama pig iron in Utah about $30.00 a ton. Chicago and St. Louis are somewhat nearer to Utah than Birmingham, Ala., but it costs $14.00 to make iron in Chicago or Missouri. It can be made in Utah, without a doubt, at $20.00, and this gives it an advantage over eastern irons of $8.00 or $10.00 per ton. What with railroad construction and operation, track repair, nine per cent of the rails must be replaced yearly. Mining, milling, fencing, building, and all the common uses of iron, there is a sufficient home market to justify an iron plant of moderate capacity in Salt Lake Valley. This market could be extended as the conditions grew more favorable, which they assuredly will do.   ·

SOME VALUABLE DEPOSITS.—The following extract from the annual report of Gov. Thomas to the Secretary of the Interior sheds new light upon some of Utah's mineral resources:

SLATE.—Deposits of slate are found in different parts of Utah. There is a very large deposit on Fremont Island in the Great Salt Lake. But the most useful and valuable discovery has been in the cañon immediately east of Provo City, Utah. The deposit of purple slate, one of the most valuable colors known to commerce, is practically inexhaustible. Samples of the slate have been sent abroad for examination and have been pronounced equal to the slate taken from the famous quarries of Wales. It has the fine grain, and the strength and durability possessed by the best roofing slate. The manufacture of shingles has commenced and the slate will now be placed upon the market for the various commercial uses.

The character of the deposit is indicated by the size of the slabs now being cut. One recently taken out measured 10x10 feet.

TRIPOLITE AND FLUORITE.—There has been discovered in the vicinity of Stockton, Tooele County, Utah, a deposit of mineral substance known as tripolite, and also a considerable quantity of the mineral fluorite or fluor-spar at Park City, Summit County, Utah.

Tripolite has been used as a polishing powder, and for this purpose goes by the name of electro-silicon. It has also been used in the manufacture of cement, in the preparation of soda silicate, and as a nonconductor of heat. But another use has been found for it by an enterprising citizen of Salt Lake City, Utah. He has turned the discovery to account by using it in the manufacture of a useful silicon soap, which seems to be growing in public favor.

The fluorite or fluor-spar has been found in the Mayflower and Anchor mines near Park City. It consists of fluorine and calcium (commonly fluoride of lime) and is white, greenish or purple in color. That from the Mayflower has all these colors. There is a sufficient quantity to encourage the manufacture of hydro-fluoric acid, used for etching glass. It might also be used in the manufacture of ornaments, and, as is sometimes done, lenses. But one of its most useful applications is the smelting of ores, fluorite being an admirable flux.

IRON ORE.—In Utah Territory iron ore can be found in all its forms. In nearly every county can be found deposits, occuring in veins, fissures and blanket, and in pockets. It is impossible to estimate the value of these deposits. Some of the ore assays as high as 60 per cent pure. In Iron County, Utah, are vast beds many miles in extent of a superior quality of hematite and magnetic ores. It is probably the most remarkable deposit of iron ore discovered in the Western World. These deposits are about 190 miles directly south of Salt Lake City, and about fifty miles from the nearest railway. Most of the ore is very pure. There are also large deposits of iron ore in Juab County. The mines of Tintic have long supplied vast quantities of ore to the smelters for use in the reduction of ores. The vast iron, coal and lime deposits of Utah will some day be utilized. When that time comes the Territory can easily supply all the iron needed by the West for many centuries.

SULPHUR.—Sulphur beds have been discovered in different parts of the Territory. The largest known as the Cove Creek sulphur mines, and situated about twenty-eight miles east of Black Rock station on the Utah Southern branch of the Union Pacific Railway, on the boundary line between Millard and Beaver Counties. The formation in which the sulphur occurs is trachyte and near the top granite. The sulphur appears in and with decomposed trachyte and volcanic tufa. The sulphur layer is from

four to twenty feet thick. Fortunately the sulphur deposit is near the surface: if it were underground the sulphurous gases would prevent it being
worked. Sulphur has been shipped from these mines for years and has
been used for selected uses.

COPPER—It has long been known that large deposits of superior
copper ore existed in different parts of Utah. There is scarcely a county
which does not contain deposits. They constitute a most important part
of the great mineral wealth of Utah.

In southern Utah a smelter is now reducing copper ores, and the matte
is hauled by wagon to the nearest railroad point fully one hundred miles
distant. This is done at a profit, and is an evidence of the rich character
of the ores. In the Henry mountains some copper veins have been discovered which abound in nuggets of almost pure copper.

UTAH ONYX.—A deposit of onyx has been found near Pelican Point,
southwest of Lehi City, Utah County. It has been determined to be composed of carbonate of calcium, commonly known as carbonate of lime.
It is, therefore, not true onyx, which is a variety of quartz, and consists
chiefly of silica.

This Utah onyx closely resembles the Mexican onyx, which is so
highly prized for decorative purposes. The deposit is reported to be about
two feet by twenty feet, and to extend downward to an unknown distance.
It is capable of receiving a very high polish and is really quite handsome.
The demand for this variety of marble, often known as "onyx marble,"
appears to be rapidly increasing for purposes of decoration, and as this
Utah onyx exhibits a greater variety of colors than the Mexican onyx it is
reasonable to conclude that if it occurs in sufficient quantity, as now seems
almost certain, it must in the near future be as eagerly sought after as
the Mexican, and will probably surpass it for all those purposes for which
the latter has been employed.

TEXTILE FABRICS.—There are a number of successful woolen mills in
the Territory, and one mill at Washington, on the Rio Virgin with cotton
machinery. Cotton is grown in Southern Utah, and a limited line of cotton
fabrics is manufactured. All the cotton used in Utah and Nevada in any form
might be grown in the warm rich valleys sloping to the Rio Colorado,
and the machinery to manufacture it be established and operated to advantage. The wool-clip amounts to ten million pounds; yet not one-
tenth of it is manufactured in the Territory. There is room and opportunity for additional woolen and cotton mills, with better machinery; and
also for cutting and making-up establishments in connection therewith.
It is a great country for hand-me-down suits; these might be made in the
towns of Utah as well as in New York and Boston.

BUILDING MATERIALS.—Raw materials in great variety have been mentioned in a previous chapter. The increase of building in the larger towns the past year has stimulated the opening of stone quarries and the establishing of brick yards. Eastern pressed brick have been sold in Salt Lake and Ogden at $72 to $82 per thousand, the purchaser paying freight on 3 tons of transportation from Philadelphia or St. Louis for every thousand brick. And this with as good clay as there is in the east (except Kaolin) abundant in Salt Lake Valley. Two or three companies are putting up machinery near Salt Lake that will supply a first-class pressed brick. It will be easy to engraft upon these establishments the requisite facilities for the manufacture of stoneware, terra cotta ware, scorifiers, fire brick—everything made out of clay except porcelain.

The Western Cement Company have within a year established a cement factory in Salt Lake, capable of making 200 barrels of 300 pounds each per day. The cement is made from a lime shale found in Parley's Cañon. It stands as much pressure per square inch, given the same time for setting, as the Portland cement, lacking about 7 per cent.

Mention has been made of the sugar-making plant which the Utah Sugar Company have put in on the sloping shore of Utah Lake between Lehi and American Fork. It involves an investment of $400,000; will eat up 350 tons of beets and turn out 40 tons of sugar per day; and will serve as an object lesson in political economy to the people who are benefitted by it—farmers and consumers of sugar.

STOCK YARDS AND PACKING HOUSE.—The Salt Lake City Union Stock Yards Company was incorporated October 7, 1890, with a capital of $250,-000 in 2500 shares, which have been taken by about 120 persons named in the incorporation, and who are residents of Salt Lake City, Denver, Omaha and City. Most of the stock is owned by railway and stock men, and of the largest stockmen of Utah, Idaho, Nevada, etc., are in the company. The directors are D. C. Dodge, manager of the R. G. W. R'y; J. W. Rodifer, investment banker, of Omaha; Geo. A. Lowe, J. E. Dooly, John H. White, W. P. Noble, R. C. Chambers, D. F. Sanders, D. M. Wells, Fred Simon, M. K. Parsons, Charles Crane, and W. H. Remington, well known citizens of Utah, representing mercantile, banking, cattle, and sheep raising and other interests.

The company have acquired 300 acres of land north of Salt Lake City, 25 of which have been put to use upon plans contemplating extensions from time to time, as may be needed. These plans include railway tracks, chutes for loading and unloading, yards, corrals, stables, sheds, etc., for horses, cattle, sheep, and hogs—in all eight divisions. There is an office and exchange building, the first floor used for the company's offices, the second for commission rooms. There is, besides, a hotel, and a weighing house. The capacity of the yards and buildings is 100 cars of cattle, 30 cars of sheep, and 20 cars of swine per day.

Connected with this enterprise are packing houses, with a capacity for cooling 500 beeves and 1,000 sheep or swine at a time, erected by White & Sons Co., of Salt Lake. The ice machine has a capacity of 75 tons of ice per day. The company have their own refrigerating cars. Much is expected from the joint undertaking in paving the way for other industries incident to this concentration of the slaughtering and curing business of the inter-mountain region.

SASH, DOOR AND BLIND FACTORY.—A company with a capital of $150,000 is to be organized in Salt Lake, for the manufacture of sash, doors, and blinds. Huttig & Co., of Muscatine, Ia,, with branches at St. Louis, Kansas City, and Wichita, are the organizers and promoters of this scheme. They will take the majority of the stock, and the chief whole-sale lumber dealers of Salt Lake will take the remainder. And thus the waste and folly of hauling lumber from California to Iowa to be made into sash, doors and blinds for Salt Lake City, which is about half way between the two, will be brought to a timely end.

Another important new enterprise is called the Salt Lake Geometrical Wood Carving and Manufacturing Company. Its capital is $300,000. It proposes to absorb the Salt Lake Mantel Manufacturing Company and the Foote Refrigerating Company, and to engage in wood carving by machines which, it is said, make their own designs, and to turn out mantels and re-frigerators and other articles in that line.

It will be seen that enterprising men are improving some of the open-ings for profitable manufacturing which for years have been advertised as abounding in Utah. However, there are plenty of opportunities left.

THE LABOR SUPPLY.—From the Governor's report we   he following:

The number of men belonging to the trades unions in Salt L. e and Ogden is as follows:

Salt Lake—
Number of trades unions federated... 3
Number of trades unions not federated...1464
Trades labor men not members of trades unions.. 872
Ogden—
Number of trades unions federated 670
Number of trades unions not federated... 498
Trades labor men not members of trades unions..
Members of trades unions outside of Salt Lake and Ogden.................................2748

Total...8835

WAGES PAID AND HOURS EMPLOYED.—The following statement will show the rate of wages paid for certain kinds of labor and the hours employed:

TRADES.	Wages per Month.	Wages per Week.	Wages per Day.	Hours Employed.
Bricklayers..	....	....	$4.50-$5.50	8-9
Brickmakers...	....	....	2.50 3.00	9
Blacksmiths	....	....	3.00 4.00	9
Boilermakers...	....	....	3.50 4.00	9
Brewers...	$ 70	....	....	9
Bakers	....	$ 21	....	9
Boot and shoe makers............................	....	....	2.75 3.00	8
Clerks...	60-100	....	....	10
Carpenters...	....	....	3.00 3.50	8-9
Cooks and waiters................................	40-75	....	....	10
Electricians	....	....	3.50 4.50	8
Harnessmakers......................................	....	....	2.25 3.00	9
Hod carriers...	....	....	2.25 2.75	8-9
Iron moulders.......................................	....	....	3.00 3.50	9
Lathers	....	....	3.00 3.50	9
Laborers..	....	....	2.00 2.50	8-9
Linemen	....	....	3.00 3.50	8
Machinists..	....	....	3.50 4.00	9
Printers...	....	21-25	....	9
Pressmen...	....	18-25	....	9
Plumbers	....	....	4.50 5.00	9
Painters...	....	....	3.00 3.50	9
Plasterers	....	....	4.00 4.50	8
Stonecutters...	....	....	4.00 4.50	8
Stonemasons	....	....	4.50 5.50	8
Steam and gas fitters............................	....	....	4.50 5.00	9
Street car employees.............................	....	....	2.00	9
Tinners and cornice makers.....................	....	....	3.50 4.00	8
Barbers	....	15-20	....	10
Cigar makers..	....	20-25	....	8

OPPORTUNITIES AND WANTS.—There are deposits of white and colored marble, of lithographic stone, of slate, of brimstone, and the materials for glass manufacture. Many of the chemical and mineral salts used in medicine and in the industrial arts are held in solution by the brine of Great Salt Lake.

Tanneries and boot and shoe factories might be multiplied by five and still fail to supply the home market.

There is demand for a great ore-treating and chemical works; one that would extract and save all the metals and useful minerals contained in any ore.

There is plenty of opportunity in the mines, the output of which might be doubled, and then doubled again.

There are unused waters to store, and the courses of unused rivers to reclaim. The facilities for water storage on the high plateaus are extraordinary, and a thousand square miles await baptism in this water on the course of the Sevier River.

There is need of men to buy the lands on the border of Great Salt Lake from Nephi to Logan, grub out the bushes, fill up the hollows, turn on the waters, and make Salt Lake Valley what it might be, but is not—the finest fruit-producing area of its size in the world. A large percentage of the fruit trees of Salt Lake Valley but cumber the ground. It is conceded that Utah has the resources of a great State, and that the people have made a fair start in their development. The output of the Territory is believed to be about $40,000,000 a year, about equally divided between agriculture, mining and manufacturing. Were the products of manufacturing multiplied by five, the output of raw materials and the production of food would increase in proportion. There is a strong call for outside men and outside money to thus increase the volume of industrial output. Why, it may be asked, do not the Utah people do this themselves? Because there is too much of it to do, and they have not the disengaged money.

The vast majority of the people of Utah confine themselves to farming on a limited scale and to the same sort of manufacturing. They have always done so, and if they had the disposition they have not the means to do otherwise. It has been left to a very decided minority to prosecute mining and its incidental pursuits. It was the promised land of the Mormons, who were segregated and gathered here from the outside, and the aim was to hold it for the Mormons rather than for other people. In the course of years, allured by the exceeding mineral richness of the country, the latter have gained a strong foothold, however, and this is now being broadened and strengthened in the most satisfactory manner by newcomers. Such money as the Gentiles of Utah have is engaged in mining, smelting, banking, merchandising, cattle and sheep raising; very little of it in agriculture and horticulture.

Many efforts have been made from time to time to start railroads, ironworks, hotels, different branches of manufactures, great irrigating schemes, and to open new mines, but unavailingly for the most part, simply from the absence of disengaged money and men to push them to success.

This is why outside men and money are called upon to bring Utah's resources into requisition. It is believed that investments in Utah at this time will prove profitable. Immigration and capital are pouring in. They must be interested in new enterprises — in railroad construction, mining, manufacturing, farming, fruit-growing, stock raising. That is why it is asserted with " damnable iteration " that opportunities exist in bewildering profusion. The coal, iron, oil and gas resources of Utah are equal, without doubt, to those of any region in the West or East; yet they are relatively untouched. Lead-silver mining and smelting are prosperous and growing, but they are still comparatively undeveloped. Especially is this true when all central and southern Nevada is regarded as a tributary field, needing nothing to make it so, in fact, but 200 or 300 miles of railroad. There would be more tanneries, foundries, boot and shoe factories, woolen and cotton mills, knitting mills, canneries, creameries, glass works, potteries, chemical factories, broom factories, powder works, but for the lack of money and men to expend it in building and carrying them on. There is scope for the use of a thousand dollars in the material development of Utah for every ten dollars now put to such use.

UNION PACIFIC RAILWAY.

PHYSICAL CONDITION.—The following memoranda on this subject was prepared by (then) General Manager Ressiguie, for Government Director Spaulding.

OREGON SHORT LINE AND UTAH NORTHERN.—The Union Pacific owns the majority of the Oregon Short Line stock. The Oregon Short Line owns the majority of the Oregon Railway Navigation stock, and guarantees six per cent on the capital.

The Oregon Short Line & Utah Northern is composed of the following roads: The Utah Division, which covers the property south of Ogden, from Oregon to Frisco, Lehi Junction to Tintic Mining Camp, and Salt Lake to the terminus of the narrow gauge road. Mileage, 380.

The Idaho Division covers the road from Ogden to Silver Bow, Granger to Huntington, Shoshone to Ketchum, Nampa to Boise. Mileage, 1,038.

The Wyoming Division, main line, covers the road from Cheyenne to Ogden. Mileage, 515.

WYOMING DIVISION.—The physical condition of the Wyoming Division is better than at any time in the history of the road.

The sidings between Cheyenne and Laramie have all been lengthened, also between Rawlins and Green River. An extensive switching yard has been built at Rock Springs, the principal mining point of the company. This yard cost, completed, $60,000, and will accommodate 1,000 cars.

Extensive yards have also been built at Ogden, which were completed at an expense of $50,000. In addition to these two main yards, smaller switch yards have been built at Medicine Bow, Rawlins and Wamsutta, and also a slight extension of the yard at Green River. There is a switch yard at Cheyenne, which cost $60,000.

IDAHO DIVISION. —The general surface of the Idaho Division track is better than ever before since the road was constructed. We were late in getting ties for the Idaho Division, on account of freshets in Oregon and heavy fall of snow in the country last winter, but the full quota of ties called for by the operating department will have been put in the track before the season closes.

There have been erected on the Idaho Division coal chutes at Fossil, Squaw Creek, Shoshone, Glenn's Ferry, Nampa, Camas, Lima, Dillon and Melrose, which aid materially in the movement of trains. These chutes cost $4,500 each. Additional tracks have been put in at Ham's Fork, Fossil, Montpelier, Manson, Soda Springs, Squaw Creek, Pocatello, Lima, Dillon and Glenn. A twenty-stall round house has been constructed at Pocatello.

UTAH DIVISION.—A new switch yard and rearrangements of tracks, and an out and in freight depot have been constructed at Salt Lake City, at a total cost of $150,000.

In addition to the improvements which have been made on the Wyoming, Idaho and Utah Divisions, the gauge has been widened between Ogden and Pocatello, 125 miles, and between Cache Junction and Preston, in the Cache Valley, a distance of fifty miles, which has obviated the transfer at Ogden and Pocatello. The new track has been laid with new sixty-pound steel and 3,000 ties to the mile, and the work has been done in a most thorough and substantial manner, and has saved on each car a haul of 240 miles.

The Union Pacific now has 606 miles of standard gauge line in Utah, and 37 miles of narrow gauge, a total of 643 miles. It has a grade thrown up between Milford and Pioche, about 140 miles. The latter is pretty sure to be completed this season, since parties engaged in mining at Pioche are badly crippled without it, and are financially able themselves to do it if need be.

THE UNION PACIFIC.—As will be seen by the accompanying map, the Union Pacific system embraces more than 6,500 miles of road, consisting of the main line, 1,037 miles long, and 5,500 miles of branches, exploring all parts of Nebraska and Colorado, extending in the inter-mountain region from Milford on the south to Butte, Mont., 700 miles, and by the Oregon Short Line, penetrating the heart of Idaho, and reaching the great and ultimate Northwest, where rolls the Columbia. The extent, variety and importance of the resources of all this vast region, opened up to the world by this great road, is illustrated by the fact that, where a few ox trains and a daily stage line twenty years ago comprised its transportation facilities, it now pays tens of thousands of dollars a day for these indispensables. Idaho has revealed a wealth of ore in the last five years which has startled the mining world, and will place her in the front rank as a mining State. The Wood River and Salmon River regions are now attracting most attention, but their prosperity has not yet fairly begun. Fertile valleys, wooded mountains, broad stock ranges, delightful resorts for tourists, and an unsurpassed climate, are Idaho's additional attractions and resources. The Oregon Short Line and Utah Northern crosses the Territory at right angles. All here said extolling Idaho might be repeated respecting Montana, and then the half not be told. The Utah Northern has had a magical effect on that magnificent Territory. Immediately it was constructed, the Union Pacific, in pursuance of its settled policy, placed before the world an attractive exhibit of her resources and attractions, to which she responded by doubling her population, wealth and production in two years. With her rich and varied mineral treasures, vast stretches of natural pasturage, fertile farming lands, and healthful climate, Montana is a most attractive, and, at the same time, practically a virgin field for capital, pluck and industry. The Union Pacific Railway system affords transportation facilities for this great inter-mountain region. It is pushing its iron way into all the valleys and over all mountains of the trade and mining centers, assisting in every way their development.

Freights received and forwarded by the Union Pacific stations in Utah during the year 1890 amounted to 997,842 tons.

As to new roads, great efforts are being made to construct a road from Salt Lake to Deep Creek. Nothing can be more important to the growth of Salt Lake City than this. It would bring under contribution—extended through Central and Southern Nevada—a larger if not better mining region than Salt Lake is now drawing upon for an output of $14,000,000 a year. It is an unexcelled mining area of great extent, stagnant from lack of railroad facilities.

GENERAL TOPICS.

POPULATION.—The population of Utah—organized as a Territory Sept. 9, 1850—by counties, by the respective censuses since taken, is shown in the following table:

COUNTIES.	†1892.	1890.	1880.	1870.	1860.	1850.	REMARKS.
Beaver.........	3,410	3,340	3,918	2,007	785		
Boxelder......	7,805	7,642	6,761	4,855	1,608	....	1880, part from Salt Lake.
Cache	16,515	15,509	12,562	8,229	2,605		
Cedar					741		Disappears; no record.
Davis.........	6,525	6,469	5,279	4,450	2,904	1,134	1880, part from Salt Lake.
Emery	3,000	4,866	556				1880, from San Pete, Sevier and Wasatch.
Garfield		2,457					1882.
Green River..					141		To Wyoming.
Grand	600	541					1890, from Emery.
Iron	2,750	2,683	4,013	2,277	1,010	300	1880, part to San Juan.
Juab............	6,200	5,582	3,474	2,034	672		
Kane	1,735	1,685	3,085	1,513			1880, part to San Juan.
Millard	4,000	4,033	3,727	2,753	715		
Morgan	1,850	1,780	1,783	1,972			
*Piute	2,200	2,842	1,651	82			1880, part to San Juan.
Rich	1,600	1,527	1,263	1,955			
Rio Virgin ...				450			1871, part to Nevada, and 1872, part to Washington.
Salt Lake	63,000	58,457	31,977	18,337	11,295	6,157	1880, to Boxelder, Davis & Weber.
San Juan......	400	365	204				1880, from Kane, Iron and Piute.
San Pete........	14,500	13,146	11,557	6,786	3,815	365	1880, to Emery, Uintah & Wasatch.
Sevier	7,200	6,199	4,457	19			1880, part to Emery.
Shamblp					162		Absorbed by Juab, Tooele & Utah.
Summit.........	8,500	7,733	4,921	2,512	198		
Tooele	4,000	3,700	4,497	2,177	1,008	152	1880, part from Salt Lake.
Uintah.........	3,100	2,292	799				1880, from San Pete and Wasatch.
Utah	27,500	23,416	17,973	12,203	8,248	2,026	1880, boundaries changed.
Wasatch	4,800	4,267	2,927	1,244			1880, to Emery and Wasatch, from San Pete.
Washington..	4,350	4,009	4,235	3,064	691	....	1872, from Rio Virgin.
Wayne.........	900						
Weber.........	27,500	23,005	12,344	7,858	3,675	1,186	1880, from Salt Lake.
TOTALS......	223,930	208,905	143,963	86,786	40,273	11,380	

*Piute County was divided by the last Legislature, the eastern part having become Wayne County.
†Estimated population.

In his report to the Secretary of the Interior for 1892, Governor Thomas says foreign immigration seems to have fallen below the usual average of other years. The domestic immigration has not equaled that of the previous year; but there has been a steady growth in the commercial and mining centers and railroad cities and towns.

TERRITORIAL REVENUE.—From the same report the appended tables are taken:

Statements showing the assessed value of the property of the incorporated cities and towns of Utah Territory, and the indebtedness of same for the years 1891 and 1892:

Cities Incorporated under Special Charter.	Assessed Value of Property. 1891.	Assessed Value of Property. 1892.	Amount of Indebtedness. 1891.	Amount of Indebtedness. 1892.
American Fork...............$	300,000.00	$ 325,000.00	$ 2,400.00	$ 2,500.00
Alpine City.........	50,000.00	40,000.00	100.00	300.00
Beaver................	280,301.00	310,412.00	244.79	911.00
Brigham City..............	464,160.00	413,410.00	....	24,000.00
Cedar City.............	145,784.00	139,868.00	....	450.00
Coalville....	215,883.80	261,287.00	....	....
Corinne.................	182,000.00	182,235.00	2,500.00	6,000.00
Ephraim	264,540.00	252,190.00	2,600.00	2,000.00
Fairview.............	120,000.00	143,200.00	....	....
Fillmore................	100,000.00	100,000.00	1,500.00	200.00
Grantsville..............	150,000.00	150,000.00	....	....
Hyrum..............	24,000.00	223,556.00	....	....
Kaysville.............	229,635.00	1,000,000.00	5,000.00	5,000.00
Lehi City..............	280,000.00	393,800.00	....	....
Logan	1,850,000.00	1,930,842.00	1,850.00	45,306.00
Manti..............................	340,000.00	362,441.50	6,000.00	12,500.00
Mendon.................	66,000.00	89,500.00	....	500.00
Moroni.................	91,284.00	95,373.00	....	....
Mount Pleasant..............	250,000.00	244,292.00	....	....
Morgan........................	207,900.00	210,000.00	....	....
Ogden	13,243,965.00	13,500,000.00	250,000.00	368,000.00
Park City...............	1,800,000.00	1,300,000.00	....	....
Parowan	108,085.00	113,950.00	217.22	146.60
Payson..............	308,500.00	323,615.00	....	· 500.00
Pleasant Grove................	350,000.00	244,030.00	....	....
Provo	3,152,620.00	8,618,646.00	1,970.00	124,000.00
Richfield..............	177,600.00	193,174.00	....	....
Richmond...............	145,000.00	156,300.00	300.00	200.00
Salt Lake City	57,965,868.00	52,598,395.00	1,000,000.00	1,500,000.00
Smithfield	192,210.00	160,000.00	537.50	....
Spanish Fork..............	237,750.00	296,230.00	....	....
Spring City...............	80,000.00	83,000.00	....	....
Springville	680,000.00	430,000.00	....	....
St. George..............	252,698.00	272,692.00	2,722.28	2,970.23
Tooele	151,742.00	160,804.00	....	....
Washington.............	42,800.82	48,700.28	....	....
Wellsville	93,000.00	113,690.00	....	....
Willard	98,986.45	102,156.75	....	....
Cities and Towns Incorporated Under the General Law—				
Bear River..............	22,430.00	30,000.00	....	....
Fountain Green.................	70,782.00	70,695.00	....	....
Heber...................	*..............	....	....	....
Kanab..............	43,600.00	53,347.00	65.00	....
Monroe	75,652.00	200,000.00	....	....
Nephi..............	823,962.00	779,854.00	16,000.00	20,000.00
Salem.............	47,317.00	144,710.00	100.00	....
‡ Salina	†..............	111,272.00	....	....
‡ Santaquin................	†..............	81,968.00	....	....
‡ Elsinore	†..............	106,450.00	....	....:
‡ Huntington	†..............	57,396.00	....	94.75
Totals...............$	55,780,856.00	$87,200,081.53	$1,294,106.79	$2,115,678.58

*No assessment.

†No assessment separate from county.

‡Not incorporated last year.

The increase in municipal indebtedness for the year is 63.3 per cent.

The increase in assessed valuation of property is 1.6 per cent.

Statement of the revenue from the tax levy for the years 1890 and 1891 for school purposes:

COUNTIES.	Territorial and School Tax. 1890.	Territorial and School Tax. 1891.
Box Elder	$ 19,847.19	$ 19,769.95
Beaver	5,024.48	5,783.53
Cache	20,014.33	30,863.60
Davis	15,813.55	17,482.47
Emery	6,474.63	7,739.62
Garfield	1,831.56	3,054.68
Grand	3,842.10	4,194.22
Iron	2,946.95	3,481.55
Juab	11,526.60	11,416 07
Kane	1,873.97	2,316.52
Morgan	4,142.00	4,606.67
Millard	3,519.70	5,914.89
Piute	3,026.93	2,444.30
Rich	3,293.65	3,829.15
Salt Lake	261,354.83	293,689.28
Summit	18,974.13	19,769.51
San Pete	14,559.43	17,298.15
Sevier	5,253.22	5,773.48
San Juan	1,382.43	1,673.36
Tooele	7,969.78	7,751.67
Utah	44,758.60	48,175.41
Uintah	2,910.23	3,071.81
Weber	73,308.78	88,412.99
Wasatch	5,430.34	5,949.12
Washington	4,011.65	4,224.19
Total	$543,061.06	$618,685.19

The increase over 1890 is 10.2 per cent. The revenue for 1892 it is estimated will be $585,754.49, a decrease of $32,930.70.

COMPARATIVE STATEMENT.—Statement showing the total revenue for each year from 1854, and the total assessed value of property from 1855:

YEAR.	Territorial and School Tax.	Value of Property Assessed.
1854	$ 6,386.31	*...............
1855	17,348.89	$ 3,469,770
1856	18,999.38	2,937,937
1857	12,892.43	2,578,486
1858	9,032.32	*...............
1859	9,957.17	3,982,869
1860	23,369.50	4,673,900
1861	25,160.92	5,032,184
1862	47,795.18	4,779,518
1863	50,482.00	5,048,200
1864	33,480.02	6,696,004
1865	47,269.65	9,453,930
1866	52,838.98	10,467,796
1867	53,239.13	10,647,826
1868	52,669.38	10,533,872
1869	59,968.03	11,393,606
1870	33,639.09	13,455,636
1871	38,163.56	15,265,424
1872	4,3976.40	17,590,560
1873	53,870.87	21,548,339
1874	57,021.45	*...............
1875	58,222.95	23,289,180
1876	50,021.11	23,608,064
1877	56,384.15	22,553,660
1878	146,903.77	24,483,957
1879	149,910.43	24,985,072
1880	151,835.24	25,222,540
1881	153,495.40	25,579,234
1882	174,483.93	29,080,656
1883	185,006.55	30,834,425
1884	203,549.64	33,924,942
1885	208,931.72	34,851,957
1886	214,105.93	35,684,322
1887	227,361.48	37,893,580
1888	282,636.61	46,868,247
1889	305,016.14	49,833,690
1890	543,061.08	108,612,216
1891	618,685.19	123,737,042

* No data from which to obtain the amount.

Statement showing the assessed valuation of real and personal property
and improvements in the several counties for 1892:

COUNTIES.	Real Property.	Improve- ments.	Personal Property.	Total. 1892.	1891.
Beaver	$ 922,276.00		$ 216,533.00	$ 1,168,809.00	$ 1,329,122.00
Box Elder	1,104,187.00	$ 370,150.00	2,559,474.00	4,033,811.00	4,094,248.00
Cache	4,044,077.00	1,027,260.00	1,273,483.00	6,344,320.00	6,158,332.00
Davis	1,989,108.00	562,095.00	1,072,444.00	3,623,641.00	3,496,435.00
Emery	261,550.00	170,255.00	564,442.00	996,247.00	1,433,786.00
Garfield	71,485.00	85,702.00	380,759.00	537,946.00	489,958.00
Grand	29,349.00	15,075.00	223,779.00	268,303.00	810,032.43
Iron	233,135.00	148,085.00	362,692.00	743,912.00	718,685.00
Juab	596,687.00	426,952.00	514,105.00	1,537,774.00	1,618,856.00
Kane	61,305.00	109,320.00	426,774.00	597,399.00	339,799.00
Millard	238,849.00	1,041,741.00	398,257.00	1,678,947.00	1,204,856.00
Morgan	319,220.00	121,470.00	164,900.00	605,590.00	907,720.00
Piute	91,114.00	35,145.00	126,439.00	252,698.00	471,180.00
Rich	488,867.00	78,635.00	223,338.00	795,778.00	796,350.00
Salt Lake	33,103,356.00	6,711,065.00	11,823,297.26	51,637,718.26	59,727,472.94
San Juan	1,600.00	800.00	361,740.00	363,940.00	334,678.00
San Pete	1,932,084.00	825,196.00	1,106,193.00	3,853,473.00	2,575,958.00
Sevier	493,817.00	251,863.00	629,198.00	1,371,875.00	1,191,915.00
Summit	1,033,771.00	1,398,938.00	1,648,906.61	4,071,615.61	3,961,593.00
Tooele	572,052.00	232,275.00	976,439.22	1,730,766.22	1,375,428.00
Uintah	180,442.00	91,588.00	296,590.00	568,625.00	629,015.00
Utah				10,244,825.00	10,357,607.00
Washington	223,375.00	251,250.00	456,248.00	930,883.00	852,226.00
Wasatch	556,405.00	255,570.00	294,985.00	1,108,960.00	1,192,730.00
Weber	10,628,143.00	3,625,777.00	3,523,311.42	17,781,231.42	18,047,000.00
Wayne	41,734.00	142,585.00	212,117.00	296,436.00	
TOTAL	$51,158,055.00	$17,885,579.00	$29,862,442.00	$117,150,899.51	$124,312,782.37

REVENUE LAW. — Section 2008 of Chapter XL., Compiled Laws, 1888,
provides for the levying of an *ad valorem* tax on all personal property as
follows: Two mills on the dollar for Territorial purposes, three mills on
the dollar for district school purposes (a Territorial tax), such sums as
the county courts of the several counties may designate for district school
purposes in such counties, not to exceed two mills on the dollar, and
such sums as the county courts of the several counties may designate for
county purposes, not to exceed three mills on the dollar.

Section 2012 provides that the property shall be assessed to the owner,
if known; if the owner be unknown then to an unknown owner. The tax
shall attach to and constitute a lien on the property assessed, if real estate,
from the 31st day of August of each year, and if personal property,
from the day of assessment. If the taxpayer owns both real estate and
other personal taxable property, the tax on personal property shall also
be a lien on real estate. In each and every case the lien is paramount to
all other liens, and it cannot be removed until the tax is paid, "or until
the title vests thereto, under a sale thereof, by virtue of proceedings to
enforce payment of the tax."

Section 2030 provides that, on or before the 1st day of December of
each year, the Collector shall publish a delinquent list, showing the
amount assessed against each delinquent in his county. This list is published for a period of ten days, and on the third Monday of December of
each year the Collector exposes for sale the delinquent property, or enough
thereof to satisfy the lien, and continues the sale from day to day until the
property of such delinquent is exhausted or the taxes and costs are paid.

Section 2032 provides for the redemption of property sold for delinquent taxes as follows: " Real estate sold for taxes, as aforesaid, may be redeemed by any person interested therein, at any time within two years after the sale thereof, by such person paying into the county treasury, for the use of the purchaser, or his legal representatives, the amount paid by such purchaser and all costs, as aforesaid (amounting in the aggregate to $7.00 on each delinquent parcel of real estate), with interest at the rate of 1½ per cent. per month on the whole from the day of sale to that of the redemption, and all taxes that have accrued thereon and which have been paid by the purchaser after his purchase to the time of redemption."

Compared with the pages of delinquent taxes published in the newspapers of "boom" towns, a pretty good showing is made for Salt Lake City and County.

EDUCATION: FREE SCHOOL SYSTEM.—On this subject the Governor has the following in his report, to wit:

The steady increase in the number of pupils attending the public schools during the year ending June, 1890, continued during the year ending June, 1891. In Salt Lake City the number of pupils seeking admission is beyond the capacity of the school buildings, and the trustees are compelled to rent private buildings.

In Ogden, Provo, Logan and other cities the schools are also crowded. The free school law has stimulated the cause of public education in every part of the Territory.

Denominational schools still exist in different parts of the Territory, though I have been informed there is a steady decrease in the number of pupils attending them. I believe it is the intention of nearly all the denominational schools to gradually withdraw from competition with the public schools.

In my last report I referred to the fact of denominational schools having been established by the Mormon Board of Education, in competition with the public schools. The statement was severely criticised by the organ of the church, and it was intimated the statement was not true. Since then I have received reports from such schools, which show conclusively that many of them are teaching the same class of studies as the public schools.

The time will soon come when the denominational school will have to give way before the public school.

DENOMINATIONAL SCHOOLS.—Statement showing the number of schools established and maintained by religious denominations, excepting the Church of Jesus Christ of Latter-Day Saints, for the years 1891 and 1892:

DENOMINATIONS.	1891			1892		
	Schools.	Teachers.	Pupils.	Schools.	Teachers.	Pupils.
Methodist	25	38	1,400	21	38	1,150
Protestant Episcopal	5	18	500	6	18	550
Catholic	6	50	800	8	73	900
Congregational	21	45	2,269	20	46	2,068
Swedish Lutheran	*	...	...	*	...	...
Baptist	*	...	...	...	...	...
Presbyterian	31	61	1,935	26	57	1,850
Total	88	212	6,904	81	230	6,518

*No report; schools discontinued.

Cost of Denominational Schools.—Statement showing the amount expended for schools by the various religious denominations, excepting the Church of Jesus Christ of Latter-Day Saints prior to June 30, 1891, and to June 30, 1892:

DENOMINATIONS.	Expended for schools to June 30, 1891.	Expended in maintaining schools, 1892.	Expended for school grounds and buildings, 1892.	Total expended for schools.
Methodist	$ 319,600	$ 11,500		$ 361,100
Protestant Episcopal	*	15,000	$ 7,000	22,000
Catholic	473,000	30,000	60,000	563,000
Congregational	386,169	35,000		421,169
Swedish Lutheran	16,500	*		16,500
Baptist			13,000	13,000
Presbyterian	374,250	27,600	2,300	404,150
Total	$1,599,519	$119,100	$82,300	$1,800,919

*No report before 1891.

The school law of March, 1890, provides for a territorial commissioner of schools, county superintendents, district trustees, and a special board for the examination of teachers in each county. All children between 6 and 18 years of age attend the public schools free of charge, and all children between 10 and 14 are obliged to attend some school, public or private, at least 16 weeks in each year, except for reasons precluding such attendance.

For the support of the system there is levied a territorial tax of 3 mills, which is apportioned by the territorial commissioner to the several counties in proportion to their school children respectively. In 1890 this tax amounted to $150,000. Each county superintendent is required to annually furnish the county court with an estimate of the money needed to carry on the schools for the year. Whereupon the county court levies such tax as may be necessary, not exceeding 2 mills on the dollar. Each district may levy a special tax not exceeding one per cent of the taxable property of the district, upon authorization by the voters, for the purpose of buldings, betterments, etc.

The districts may borrow on 6 per cent 5-20 bonds; districts of less than 500 inhabitants may thus borrow $3,000; of more than that number, up to 2 per cent. of the assessed valuation. Interest on these bonds is raised by a tax not exceeding 2 mills on the dollar, and after 5 years a similar tax may be levied to meet the principal.

The cities are divided into two classes, in both of which each city forms a single school district by itself. There is a board of education composed of the mayor and two trustees from each municipal ward. All the school property of the city vests in the board of education. The board may fill vacancies between elections. It appoints a superintendent, clerk, and an examining board.

DESERET UNIVERSITY.—The Territorial (Utah) University, at Salt Lake, under the fostering care of the legislature, and managed by an eminent and capable board of regents, is fast becoming the favorite educational institution of Utah and the intermountain region. The board of instruction contains a score of teachers, and it is at present organized with seven departments. While the courses are to a considerable extent elective, all require, as a basis for advancement, a thorough knowledge of the branches necessary to a good business education, and in the higher courses the university is prepared to give as complete an education as can be acquired in eastern colleges. In many specialties, such as music, art and mechanical drawing, painting etc., it affords a wider range of instruction than most colleges. The legislature of 1890 established a department of geology and mining, and this chair is filled by a professor of high literary and scientific attainments, who brings to his work several years of successful experience. He is also curator of the museum, the extent and scientific value of which is already considerable. The university has a library of 10,000 volumes, which will be added to as means in hand permit. A military class is maintained as a substitute for gymnastics in giving regular moderate physical exercise, a manly deportment, and teaching habits of promptness and attention. Spanish, in addition to German and French, is taught, with a view to the future commercial relations of this country with the lands to the south, of which Spanish is the language. One of the most important features of the university is its normal department. It is obliged by law to teach a perpetual class of 100 teachers. The state that provides free common schools, and competent teachers for them, has done its duty in the matter of education. Ninety thousand dollars was appropriated for the University for 1892 and 1893.

The people of the intermountain region are so accustomed to look abroad for instruction in the higher educational branches, that the first and perhaps the most difficult lesson the university will have to teach is, that what students seek abroad they can get at home at half the cost; but with the known healthfulness of the climate, the moderate cost of board in Salt Lake City, the nominal charge for tuition, the liberal support of the legislature, and the confidence and encouragement of the people, supplemented by a progressive management, keeping pace with the development of the Territory, the university in the near future should attain a high degree of prosperity and usefulness, and a reputation commensurate with its merits.

The university has a deaf mute department, handsomely housed in University Square, and it is the only school of the kind between the Rocky Mountains and the Pacific Coast. It is already receiving students from Idaho and Arizona. The attendance in 1890 was 43. During the year an industrial department was added to the school: printing, carpentering, and shoemaking for the boys, and sewing and fancy work for the girls. Some of the work done was on exhibition at the Territorial fair of last fall, and received a silver medal and a diploma.

CHURCHES.

Statement showing the number of churches and ministers maintained by religious denominations, excepting the Church of Jesus Christ of Latter-Day Saints, for the years 1891 and 1892:

DENOMINATIONS.	1891.		1892.	
	Churches.	Ministers.	Churches.	Ministers.
Methodist	33	26	35	30
Protestant Episcopal	8	7	10	8
Catholic	6	15	9	19
Congregational	5	8	8	12
Swedish Lutheran	6	4	†6	7
Baptist	*	...	3	4
Presbyterian	17	19	18	20
Total	75	79	89	98

*No report. †Also 10 missions without churches.

COST OF CHURCHES.—Statement showing the amount expended for churches by the various denominations, excepting the Church of Jesus Christ of Latter-Day Saints, prior to June 30, 1891, and to June 30, 1892:

DENOMINATIONS.	Expended for grounds, buildings and maintenance.		
	To June 30, 1891.	To June 30, 1892.	Total.
Methodist	$217,500	$ 8,500	$226,000
Protestant Episcopal	*.........	1,200	1,200
Catholic	113,000	7,000	120,000
Congregational	20,000	51,000	71,000
Swedish Lutheran	45,900	6,275	52,175
Baptist	*..........	*...........	....
Presbyterian	89,700	2,300	92,000
Total	$486,100	$ 76,275	$562,375

*No report.

SETTLEMENT OF PUBLIC LANDS.—Statement showing the disposition and settlement of public lands in Utah Territory, and the total business of the Land Office at Salt Lake City from the time of its opening in March, 1869, to the end of the fiscal year ending June 30, 1892:

	No.	Acreage.	Amount.
Cash entries	4,008	378,843.52	$ 596,816.17
Mineral entries	1,840	20,063.99	96,467.00
Mineral applications	2,137	10,987.77	21,370.00
Desert applications	3,565	716,387.30	187,184.21
Desert final entries	856	158,709.35	161,429.24
Homestead entries	9,805	1,233,966.24	150,281.65
Homestead final entries	4,768	673,549.70	31,315.85
Timber culture entries	1,577	179,303.49	17,902.00
Timber culture final entries	18	1,900.00	72.00
Adverse mining claims	926	...	9,026.00
Pre-emption filings	11,993	1,444,727.88	35,979.00
Coal filings	1,144	144,120.00	3,432.00
Coal cash entries	105	189,933.80	13,340.20
U. P. and C. P. R. R. selections	......	839,068.30	8,039.64
Land warrants	......	23,957.00	615.00
Agricultural College scrip	......	84,912.00	2,232.00
Valentine scrip	......	280.12	14 00
Chippewa scrip	......	479.82	10.00
Supreme Court scrip	......	4,530.02	....
Sioux Half-breed scrip	......	360.00	6.00
Timber sold	......	....	127.08
*Timber depredations	......	....	15,422.31
Testimony fees	......	....	17,142.92
Total	......	5,906,080.30	$1,368,224.17

*Timber depredations and stumpage consolidated.

Statement of the business of the United States Land Office at Salt Lake City, Utah, for the fiscal year ending June 30, 1892:

KIND OF ENTRY.	No.	Acreage.	Amount.
Cash entries (including acreage in commuted H. E. & T. C. E.)	187	12,230.50	$26,449.52
Mineral entries	89	*2,015.76	9,520.00
Mineral applications	125	2,442.18	1,250.00
Desert applications	224	33,903.55	8,786.46
Desert final entries	87	*23,460.79	25,379.49
Homestead entries	637	87,569.24	9,617.94
Homestead final entries	190	*26,740.40	1,201.22
Timber culture final entries	3	*320.00	12.00
Adverse mining claims	38	....	380.00
Preemption filings	4	*385.18	12.00
Coal filings	39	*5,480.00	117.00
Coal entries	8	1,195.94	22,318.80
Railroad selections	577	92,319.65	1,153.00
Testimony fees	......	....	783.64
Total	2,208	229,666.06	$106,986.07

*Not new entries.

Total area surveyed to June 30, 1891, 13,198,204.16 acres.

BUSINESS, BUILDING, SALT LAKE VALLEY.—In former times Utah took no part in the systematic advertising and colonizing which have so hastened the settlement of the West. Little was heard of the relative advantages of Utah as compared with her neighbors. Her resources, improvements, wealth, industries, population, position, climate, attractions, capabilities and opportunities were not dilated upon. About six years ago there was a sudden change in this respect. Aspiration for the material advancement of the Territory seemed to seize all classes, and from that moment dawned a new era for Utah. The spirit of enterprise has for the past three years hourly grown stronger as it has materialized in new railroad lines, new buildings, public and private, new factories, new industries, new mines, rapid transit, electric lines, sewers, pavements, sidewalks, improved lighting and agricultural development of the Territory. During that period the city and Territory have opened and expanded like a vast flower. Faith and courage have been richly rewarded, and to realize the most sanguine dream of to-day nothing is requisite but the development of these qualities without hesitation or reserve. To the invalid, to him who is seeking an ideal climate, to the business man or the manufacturer who is seeking a new location, or to the investor, Salt Lake Valley is without doubt the most inviting of all the flowery fields which to-day are attracting attention.

As if to forever bar a water famine in Salt Lake Valley, nature has provided a reservoir in Bear Lake, 150 square miles in area, high up in the mountains on Bear River, the principal tributary of Great Salt Lake from 'the North; and a second reservoir in Utah Lake, 125 square miles in area, on the Jordan, the principal tributary of the Lake from the South. There is thus, with Weber River, entering the valley and the lake from the East, water enough forever assured to irrigate every acre of the eastern border of Great Salt Lake, from Nephi on the south to Bear River Cañon on the north, a distance, as traveled, of about 150 miles. This fringe of the desert, between the Wasatch and Great Salt Lake, and between the Wasatch and Utah Lake, about 1,200,000 acres in extent, is potentially, in location, resources, climate and fertility, the glory of the earth. It is easily the garden spot of Utah. It supports more than 30 settlements or towns, and more than half the population of Utah. Every acre of the land is worth $100, although it varies in price, exclusive of the suburbs of the larger towns, from $5 to $225 an acre. Two railroads extend from end to end of it, and thence to all the world; hardly an acre of it is more than five miles from these roads; there is a perennial flow from the overshadowing great mountains of 10,000 second-cubic feet of water, sufficient to irrigate the .whole. expanse; there is a cash market at good prices in the adjacent mines, and in the trading, manufacturing and professional population of the towns. There is nothing like it between the eastern half of Nebraska and the valley of the Sacramento.

Salt Lake Valley has reason to be satisfied with its central and commanding position, whether with respect to the belt of fertile land sheltered and watered by the cloud-compelling mountains, and already planted with 200 towns, and settlements, stretching from the Rio Colorado on the south to the Snake River on the north; or with respect to the vast and virgin region, rich in all sorts of resources, and especially in mineral resources, sweeping away beyond the encircling horizon, and bounded on the southwest only by the Pacific Ocean. It is the industrial, the commercial, the business heart, not only of Utah, but of the whole imperial expanse lying between the crest of the Rocky Mountains and the crest of the Sierra Nevada, and extending from the Main Range dividing Montana and Idaho on the north to the Colorado River on the south. The most extravagant present estimate of the resources and possibilities of this empire will be more than realized in the future.

And, by the way, Salt Lake City is the capital of Salt Lake Valley, with all that the term implies.

The extent and value of new buildings throughout the Territory is almost beyond computation; certainly no conclusive figures are attainable.

Under this heading the Governor's report says:

BUSINESS PROSPERITY.—During the year ending June 30, 1892, there has been a steady development of the. business interests of the Territory, though not to the same extent as in the years 1890 and 1891. In the commercial centers business has been quiet, but in the remoter counties many of the new settlements have been growing quite rapidly.

In the principal cities and towns the population has steadily increased, and the number of persons coming to the Territory from the East is quite large. The statistics show that new buildings have been erected in the different cities and towns to the value of $3,017,380.00.

The sugar manufactory erected at Lehi, Utah County, is now in successful operation. Because of its presence the price of sugar was lower in 1891 than it has been in the history of the Territory.

NEW BUILDINGS.—Statement showing the number of residences and business buildings erected or under contract for erection in the cities and towns for the year ending June 30, 1892:

Cities Incorporated Under Special Charter.	Dwellings.	Value.	Business Houses.	Value.
American Fork	26	$ 16,600	4	$ 6,650
Alpine	3	3,000	...	....
Beaver	10	12,000	2	5,000
Brigham City	28	32,000	3	10,750
Cedar City	10	10,500	3	1,200
Coalville	9	4,500	...	....
Corinne	3	5,000	2	18,000
Ephraim	10	9,700	3	35,000
Fairview	10	3,000	2	500
Fillmore	2	5,000	2	1,000
Grantsville	....	....	....	....
Hyrum	5	4,000	1	2,000
Kaysville	11	15,000	...	....
Lehi	40	35,000	3	22,000
Logan	66	81,700	4	43,300
Manti	20	17,800	3	4,900
Mendon	6	3,500	...	....
Moroni	10	5,000	...	....
†Mount Pleasant	23	16,875	1	300
Morgan	6	6,000	...	....
‡Ogden	165	296,210	29	281,500
Park City	65	45,500	3	83,000
Parowan	1	1,500	...	....
Payson	30	24,000	4	15,000
Pleasant Grove	10	6,000	4	7,000
§Provo	32	101,000	5	15,800
Richfield	6	5,000	15	18,025
Richmond	5	7,500	...	....
Salt Lake City	545	952,294	39	585,775
*Smithfield	6	6,000	...	....
Spanish Fork	12	11,900	4	17,000
Spring City	12	5,000	2	1,400
Springville	32	32,000	7	56,000
St. George	6	9,000	...	....
Tooele	2	2,000	1	8,000
Wellsville	3	2,700	...	....
Willard	4	5,000	1	3,500
Washington	....	....	...	....
Cities and Towns Incorporated Under the General Law.				
Bear River	4	5,000	...	....
Monroe	8	4,123	...	....
Fountain Green	2	2,500	...	....
Heber	5	6,000	2	5,000
Kanab	4	2,000	1	3,000
**Nephi	....	....	...	....
Salem	7	3,000	...	....
Salina	7	5,500	12	13,000
Santaquin	9	4,600	3	8,000
Elsinore	13	6,300	8	7,100
Huntington	3	3,500	...	....
Totals	1,296	$1,827,384	171	$1,190,000

*District school, $1,600. †District schools, $10,000. ‡Public schools, $220,000. §Public schools, $16,500; Brigham Young College, $75,000. **No report.

SALT LAKE CITY—BUILDING IMPROVEMENTS.—Two years ago the fire department consisted of eight men, one steamer, one hose cart, two stations. There are now 65 men—25 full paid, 20 call men paid $50 a year, 20 volunteers, 1 life-saving corps; 2 Silsby engines, 6 hose carts and reels, 2 hook and ladder trucks, 1 chemical engine, 6 stations.

Sixty-five miles of electric or motor street car lines have been completed. There are six street car lines, namely: The Salt Lake City, the Rapid Transit, the Great Salt Lake, the West Side Rapid Transit, Beck's Hot Springs, and the East Bench. These lines cover the principal streets of the city, and extend four or five miles from the intersection of Main Street and First South in every direction. The electric light works have established 250 arc lights at the intersection of the streets, and put in an incandescent lighting plant costing $165,000. Residences are furnished with electric lights cheaper than in Denver or San Francisco.

There are sixteen banks in the city—six national, seven incorporated under Territorial law, and three private banks. The six national and seven incorporated banks report a combined capital of $3,058,082, $472,412 surplus, $323,028 undivided profits, $5,725,301 deposits, and $5,805,704 loans and discounts. Add to these the estimated capital, deposits and loans of three private banks, to-wit: $1,000,000 capital, $2,500,000 deposits, and $2,250,000 loans, and the showing for Salt Lake is, banking capital, surplus and undivided profits, $4,853,522; deposits, $8,225,301; loans and discounts, $8,055,704. The laws allow national banks to loan all their capital and from 75 to 85 per cent of their deposits. In Salt Lake, if all the capital, surplus and undivided profits were loaned, then but about 40 per cent of the deposits are loaned. The Salt Lake Clearing House Association began operations April 1, 1890. The clearings for 1891 and 1892 were as follows:

Month.	1891.	1892.
January	$ 8,776,471	$ 7,587,452
February	7,292,928	6,238,626
March	6,162,690	7,461,484
April	7,128,929	9,006,519
May	5,427,098	8,374,002
June	5,821,944	7,971,650
July	7,200,625	8,363,500
August	7,949,917	7,493,757
September	6,037,262	7,152,292
October	6,649,649	7,818,726
November	6,972,030	9,481,017
December	7,413,277	8,064,589
Totals	$81,783,820	$94,023,611
Increase		$12,239,791

The exchange for January 1—$1,770,977—exceeded that of twenty-one cities on the list in the United States; it was one-fourth that of Montreal, of Kansas City, of Milwaukee, Buffalo, Galveston and Minneapolis; and they have steadily increased up to the present time.

The business of the post office, of the telephone exchange, of the telegraph office, and the sales of real estate, show an increase of 40 to 50 per cent for 1889 upon the business for the preceding year.

The assessed value of property in the city shows a steady advance; not a boom, but something substantial.

OGDEN.—The following is from an account published in 1890. Of course the city has advanced greatly since then:

"The city is nestled up against the towering peaks of the Wasatch Range on the western slope, and from which the fall in elevation is gradual and easy down to the Weber and Ogden rivers, which unite at a point west of the city and flow across the world-famed valley into the Great Salt Lake. The city is built upon the triangle formed by the junction of these two rivers. The view obtained of these rivers, the Great Salt Lake and the lovely green valley stretching as far as the eye can reach north and south, from almost any portion of the city east of the main thoroughfare known as Washington Avenue, is a grand one, and leaves an impression upon the mind while memory lasts. Where once was naught but a vast and barren waste, to-day stands Ogden, the Queen City of the West, with a population of nearly 18,000 souls, and around it gleams and shimmers wide fields that grow the golden grain, beautiful orchards of delicious fruit, and thousands of comfortable homes.

"Ogden is situated midway between Omaha and San Francisco, in the heart of one of the largest and most fertile agricultural and horticultural regions between the Missouri and the Pacific. It is in the heart of a mineral region that is just beginning to be developed; it is situated in one of the most beautiful and healthful portions of the globe, and is surrounded by regions of unsurpassed scenic and historic interest. The city is situated in one of the greatest sanitariums in the world, abounding in facilities for lake bathing, mineral springs of every known character, the purest mountain air, and the most healthful climate. Five lines of railways branch out in all directions, and others are pushing westward to this point.

"Within the territory tributary to Ogden lie more than 1,000,000 acres of tillable land unexcelled in fertility, and not even one failure of crops is known. Vast storehouses of iron, coal, limestone, salt, oil, soda, asphalt, natural gas, slate, marble, building stone and fire clay, as well as mountains of gold, silver, lead, copper and other precious minerals, are at its doors.

"At the present time it is almost impossible to find a residence or business house for rent; buildings cannot be constructed with rapidity enough to supply the demands. There are twenty-five wholesale and twenty-two manufacturing establishments located here, and every business house and factory is doing a thriving business. There has not been a business failure of importance during the last five years, and but forty in many years. It is a fact that during forty-two years there has never been a tax deed recorded, and it is equally remarkable that, in the history of the county, but four mortgages have been foreclosed. Artesian wells are found and flow abundantly at a depth of one hundred feet. Ogden has every facility for profitable manufacturing enterprises, namely: a large supply of raw products in every variety, a large and rapidly growing market, from which outside competition can be excluded, cheap fuel and water power, and unequalled transportation facilities.

"The assessed value of Ogden real estate has increased in the past year nearly three hundred per cent., and the aggregate sales in the past year have reached nearly $7,000,000, which is nearly $1,000,000 per month during the term of real estate activity.

"In 1888 Ogden expended $300,000 in new buildings, in 1889 the expenditure amounted to $1,451,727, and in 1890 her new buildings will foot up at the least $3,000,000, an increase over last year of $1,548,273, and over the previous year of $2,900,000; which is a wonderful showing, and is due to the natural advantages, railway facilities, and the courage, progressiveness and enterprise of the business men. The wealth of the city has increased during the year about $18,000,000.

"In the way of public improvements Ogden has done remarkably well for a city of its size. Something like twenty miles of asphalt side-walks have been laid; two systems of electric lights are in operation; two systems of electric railways are about completed, and are partly in operation; they together constitute about thirty-five miles of tracks. The streets have been macadamized and improved; a sewerage system is under construction, and will be ready for use by spring; the fire department is one of the finest in the Territory; and the public school system is one of which the people can well feel proud. Ogden is the seat of the Utah Wesleyan University.

"The city is well supplied with churches, built in modern styles, which are located in desirable places. The attendance is large in all of them, and the denominations are: Later-day Saints, Congregational, Methodist, Baptist, Episcopal, Presbyterian and Catholic.

"As to railroads, Ogden is the terminus of three main trunk lines, viz.: the Union Pacific, Southern Pacific and Rio Grande Western systems, with branches belonging to the former.

"Civic societies, such as the Masons, Knights of Pythias, Independent Order of Odd Fellows, Ancient Order United Workmen. Catholic Knights, Patriotic Order Sons of America, and Royal Arcanum, are in prosperous and healthy condition. The city is well supplied with newspapers, having two daily with an equal number of weekly publications, as well as others conducted by certain schools and churches.

"Natural gas in great quantity has recently been struck in the western suburbs of the city, and already the enterprising citizens are claiming an increase of population to 100,000, within the next five years with wealth in a still greater ratio. The growth of all the towns of the Territory, while not so vigorous as that of Salt Lake City and Ogden, was very satisfactory; of many of them it was extraordinary."

Ogden has two daily papers—the *Standard* and *Post*—the former edited by Frank J. Cannon, a son of President George A. Cannon, of the Mormon Church, and recently the Republican candidate for Delegate to Congress; the latter by A. L. Rhodes.

BANKS AND BANKING.

The following statement of business of the banks of the Territory, made on the 30th of June last, is official:

NAME.	Capital.		Deposits.	
	1891.	1892.	1891.	1892.
BRIGHAM CITY—				
Bank of Brigham, branch Ogden, Utah, L. & T. Co	$ 25,000.00	$ 85,000.00	$ 61,275.48	$ 61,324.15
Bank of Spanish Fork	*	19,425.00		10,550.79
CORINNE—				
J. W. Guthrie	55,000.00	50,000.00	20,000.00	65,000.00
Davis County Bank	*	11,706.50		5,423.26
KAYSVILLE—				
Barnes Banking Co	25,000.00	25,000.00	25,367.15	43,045.76
LEHI—				
Commercial & Savings Bank	*	49,000.00		46,832.62
LOGAN—				
Thatcher Bros. Banking Co	150,000.00	150,000.00	162,821.21	193,658.00
MANTI—				
Manti City Savings Bank	25,000.00	25,000.00	78,396.07	112,328.17
MT. PLEASANT—				
Mt. Pleasant Commercial & Savings Bank	*			
NEPHI—				
Savings Bank & Trust Co	50,000.00	50,000.00	48,184,46	43,625.50
First National Bank	80,000.00	50,000.00	117,861.24	113,330.74
OGDEN—				
Ogden State Bank	138,000.00	125,000.00	105,000.00	145,000.00
Commercial National Bank	150,000.00	150,000.00	230,000.00	233,302.06
First National Bank	75,000.00	150,000.00	191,295.00	396,467.52
Utah National Bank	200,000.00	100,000.00	325,000.00	300,000.00
Citizens' Bank	145,290.00	150,000.00	113,364.52	232,873.50
Ogden Savings Bank	75,000.00	75,000.00	177,365.41	217,229.45
Utah Loan & Trust Co.'s Bank	200,000.00	215,000.00	91,033.30	108,437.09
PARK CITY—				
Park City Bank	50,000.00	50,000.00	88,127.06	155,022.04
PAYSON—				
Exchange & Savings Bank	25,000.00	30,200.00	26,443.14	59,781.62
PROVO—				
Commercial & Savings Bank	75,000.00	75,000.00	68,066.00	75,015.00
First National Bank	50,000.00	50,000.00	57,563.47	43,563.00
National Bank of Commerce	53,654.28	50,000.00	27,230.96	27,755.58
Utah County Savings Bank	50,000.00	50,000.00	52,553.67	50,000.00
RICHFIELD—				
James M. Peterson	20,000.00	20,000.00	22,660.65	26,286.00
SPRINGVILLE—				
Springville Banking Co	*	50,000.00		29,583.09
SALT LAKE CITY—				
American National Bank	265,000.00	250,000.00	296,222.53	404,423.41
Commercial National Bank	330,000.00	310,000.00	334,469.82	459,767.81
Deseret National Bank	500,000.00	500,000.00	841,073.00	1,153,200.64
National Bank of the Republic	505,000.00	500,000.00	331,488.44	332,785.00
Union National Bank	440,000.00	445,000.00	908,834.17	847,408.50
Bank of Commerce	100,000.00	100,000.00	162,948.89	260,209.40
State Bank of Utah	500,000.00	500,000.00	250,286.31	422,948.30
Deseret Savings Bank	100,000.00	100,000.00	424,941.23	557,886.50
Zion's Saving Bank & Trust Co	127,287.00	400,000.00	927,596.46	1,033,496.24
Salt Lake Valley Loan & Trust Co		200,000.00	†	
Utah Title, Ins. & Trust Co. Savings Bank	160,000.00	150,000.00	77,725.64	151,220.65
Wells, Fargo & Co	200,000.00	200,000.00	1,324,940.63	1,330,940.05
W. S. McCormick & Co		200,000.00		1,200,000.00
T. R. Jones & Co				330,013.11
Utah Commercial & Savings Bank	200,000.00	200,000.00	240,272.00	333,500.44
Utah National Bank	200,000.00	200,000.00	150,235.47	290,445.24
TOTAL	$5,148,231.78	$5,910,331.50	$8,355,584.39	$11,913,750.17

*New bank.   †Receive no deposits.

Increase in bank capital, 16.7 per cent; increase in deposits, 42.1 per cent.

PUBLIC BUILDINGS.

The Capitol building erected in Fillmore many years ago is now occupied by the public schools of that city.

The new buildings erected at the penitentiary grounds are now in use and seem to be adequate to the wants of the institution.

The Industrial Home is under the control of the Utah Commission, who are required by law to make an annual report to Congress.

At the last session of the Legislative Assembly $40,000 was appropriated for the maintenance of the Reform School for the years 1892 and 1893, and $50,000 for new buildings.

It is reported that the growth of the Agricultural College exceeds that of any like institution in the West. The Legislature appropriated $65,000 to complete the buildings in accordance with the plans originally adopted. There is now in attendance some 225 pupils, from eight States and Territories.

The Territorial Capitol site consists of 20 acres on the North Bench, head of First East Temple Street, Salt Lake. A building to cost $3,000,000 is contemplated. The Capitol Commission have provided water, fenced the ground, and put out trees. The last legislature appropriated $10,000 for the care and improvement of the grounds. The erection of the building was laid on the table.

The Insane Asylum is at Provo. The building has cost $288,000, and ranks in completeness with any in the country.

Ogden has erected a new and spacious City Hall, and Salt Lake City and County are jointly engaged in the construction of a City Hall and Court House combined, to cost $300,000.

A branch of the Keeley Institute was established in Salt Lake City on January 10, 1892, with a corps of physicians. Temporary quarters have been secured in the famed "Gardo House," but the Company announces that it will soon have a building of its own to cost $50,000.

ATTRACTIONS.

Salt Lake City.—Among pleasure resorts in Utah, Salt Lake City ranks first. Laid out on the alluvial cone of a mountain stream which tips it gently up toward the setting winter sun; sheltered on the east and north by the towering Wasatch; with the beautiful Jordan Valley unrolled in its front and Great Salt Lake within cannon shot; its spacious streets bordered by trees and singing brooks, and far enough apart to give ample room for buildings, gardens, orchards, and ornamental grounds; enjoying the most agreeable and healthful climate of perhaps any large town in the United States, with ample hotel accomodations, electric-lighted streets and houses, good water, street cars propelled by electricity, thermal springs, a cheap and abundant market, churches of the principal denominations and good schools, live newspapers, telegraph lines and railroads and telephones, public parks, libraries, theatres, hospitals, fine drives and fine stock, trout fishing in the adjacent cañons and duck shooting within easy reach; withal the Mormon Temple and Tabernacle, the Deseret Museum and Salt Lake Chamber of Commerce, and the certainty that the town is to double in wealth and population in the next five years—such is in few words Salt Lake City.

From the city one can visit Great Salt Lake, the Cottonwood mines, the Cottonwood lakes, the Bingham mines, the Tintic and the Stockton mines; the Park City and the American Fork mines, and return the same day if he chooses. The Warm Springs are piped into the heart of the city and poured perpetually through a great swimming pool, piping hot. The Hot Springs are ten minutes out by rail. Great Salt Lake is reached on the south and the east shore—at Garfield and Lake Park—in twenty miles. The mining town of Alta, in Little Cottonwood, is twenty-five miles out, whence an hour or two on horse-back brings one to the Big Cottonwood lakes, to the American Fork mines, to Parley's Park and Park City. Bingham is the same distance from the city by rail as Alta; Tintic and Stockton twice as far. American Fork Cañon is taken by wagon from the town of that name, thirty-two miles south of the city. An interesting point three miles from Main Street is Fort Douglas, a well built, full regiment post, on the east and 500 feet above the city. The post and grounds are laid out with taste, water from Red Butte Cañon making cultivation possible. The elevation affords a fine view of valley, city, and lake. The latter lies a blue band along the base of island mountains in the north west, the vistas between which are closed by mountains behind mountains. In the north the Promotory divides the waters, Across Jordan Valley the horizon rests on the snow-caps of the Oquirrh. On the south the opposing mountains clasp hands, shutting in the valley in that direction.

Amongst the most attractive objects in the city are the Tabernacle—built for use, not grace—with a seating capacity of 10,000, and a fine organ; the granite Temple, 100 by 200 feet on the ground, and 100 feet to the base of the towers completed externally and to be finished internally by April next; the Salt Lake Museum, a valuable collection of Utah minerals and antiquities; Salt Lake Chamber of Commerce, containing many of the products of the farms, the gardens, the mines, mills and factories of the Territory; the Exposition building; the Eagle gate, built by Brigham Young; the Hot Springs and the Warm Springs, with all kinds of baths and conveniences; the tower on Prospect Hill, Liberty Park, the boulevard, the farms, gardens, orchards, and meadows in the suburbs. There are good public buildings and many noble residences, with well-kept grounds. Steam cars and street cars run hourly to Fort Douglas and the nearest cañons; also to the Hot Springs, and steam cars to Great Salt Lake, and to Park City *via* Parley's Cañon. The city has ever in view the magnificent chain of the Wasatch, which rises abruptly to a height of 8,000 feet above the valley, with no foot-hills to dwarf their proportions. Much of the year they are white with snow. In the autumn they wear all the colors of the rainbow in succession, as their grasses and shrubs are touched more severely by the frosts. In the spring their lower slopes take on a shade of green. On northern exposures they are dark with pines. Their general summer hue is gray, although their light and shade and color are as changeable as the winds that play about their craggy crests, invade their recesses, and in their untiring movements have chiseled gorges in the solid rock thousands of feet deep, giving infinite variety of profile and contour. The geologists tell us that as the Wasatch and the Uintahs were raised up out of the seas, ten miles were removed from their rising crests by erosion. The Wasatch are seen to the best advantage from Salt Lake City. The Twin Peaks, overshadowing Jordan Valley, have an absolute altitude of 12,000 feet, and the peak further south is 500 feet higher still.

MINERAL SPRINGS.—Of the chemical and thermal, salt, sulphur, soda and chalybeate springs, which occur in different parts of Utah, the Warm Springs and the Hot Springs in the suburbs of Salt Lake, and the Utah Hot Springs ten miles north of Ogden, are best known, best improved and most resorted to for recreation. Of the Warm Springs, Dr. Jackson, of Boston, says the water is a Harrowgate water, abounding in sulphur. It is very limpid, has a strong smell of sulphuretted hydrogen, and contains the gas, both absorbed in the water and combined with mineral bases. It is slightly charged with hydro-sulphuric acid gas; and is a pleasant saline mineral water, having the valuable properties belonging to saline sulphur springs. Issuing from under the mountain in large volume—temperature 95° to 104°—the water is conveyed in pipes into two or three bathing houses, containing plunge, shower, and tub baths, and dressing and waiting rooms. The property is owned by the city, but is under lease for a term of years to parties who are bound by the terms of the lease to make extensive improvements, and who are doing so at this writing. The cars (propelled by electricity) run out there, and the springs are visited by everybody, the water being considered efficacious in the alleviation or cure of paralytic, rheumatic, scrofulous, and other diseases.

The Hot Springs are further out—about four miles from the post-office. The water boils out from under the massive rock foundation of the mountains in such volume as to be the cause of a lake covering 1,200 acres. The temperature is 128° and the sulphurous fumes fairly stifling. The water is very similar to that of the Warm Springs, yet bears an even higher repute. It is alleged to exceed in curative properties that of the Hot Springs of Arkansas. There are all kinds of conveniences for bathing—hot, cold, tepid, single, double, and swimming or plunge baths —and limited boarding accommodations for visitors. The proprietors are constantly improving the place. There is no use in wasting words on these springs. They are convenient to the town, and can easily be tested. Indeed, as has been said, they are piped into an immense natatorium on West Temple Street within half a block of the Continental Hotel. The company which did this contemplate piping the waters of the Great Salt Lake into the city for another natatorium.

UTAH HOT SPRINGS.—What were long known as the Red Springs, ten miles north of Ogden, are hot water so impregnated with iron as to kill the vegetation over a large area and color the ground red. A large building for the use of these springs in any way experience may suggest, chiefly at present for bathing, was erected in 1878. This has since been supplemented by other improvements until the Utah Hot Springs, as they are now called, are the best equipped of any in the West. The waters pour forth in great volume from crevices in the rocks, at a temperature of 125°, and contain such ingredients as chloride of sodium, iron, magnesia, and nitre, in strong solution. For years the waters of these springs have been known to possess peculiar medicinal properties. In early days the people for miles around would come and carry away the water in barrels and casks, and it would be used as a blood purifier.

For rheumatic and some kidney troubles, nothing can surpass the waters of the Utah Hot Springs. Men of eminence who have personally tested them have said to the writer, that if the remedial effects of the use of these waters in rheumatism could be made thoroughly and widely known, people would visit them by train loads from all over the United States. Although the water is quite salt, the mixture of other mineral solutions it contains makes it drinkable, and it is used internally with advantage. The springs are a station on the Union Pacific line from Utah to Montana, and trains pass up and down every day, making them easy of access. A motor street railway connects them with Ogden, and upon this trains run hourly or oftener.

Further north, twelve miles from Bear River Cañon, is a group of springs issuing from between strata of conglomerate and limestone, within a few feet of each other, of which one is a hot sulphur, a second warm salt, and the third cool drinkable water. The volume from these springs is copious, but they run some distance before they become thoroughly mixed, although in the same channel.

HOTEL ACCOMMODATIONS.—These in Salt Lake City were regarded as fairly good and ample, before the Templeton, Knutsford, Morgan, Union Pacific, costing together nearly $1,000,000, were added. Besides these, a number of buildings were built for or transformed into boarding houses, and the upper floors of others for lodgers. In this list should not be overlooked the once noted Continental, formerly Townsend House, the first hotel of the first rank in the city; it is to be demolished and a great structure take its place. There are a half dozen first-class hotels in the city, and as many more called second-class, but whose rooms, furnishing and tables are second to the first-class only in prices, and perhaps in style. Ogden has added to the Broom, the Reed Hotel—a fine large house, first-class in all its appointments. Provo, Payson, Logan, and other towns of their class, have comfortable houses for the accommodation of visitors.

GREAT SALT LAKE.—The first mention of Great Salt Lake was by the Baron La Hontan, in 1689, who gathered from the Western Indians some vague notions of its existence. He romanced at length of the Tahuglauk, numerous as the leaves of the trees, dwelling on its fertile shores and navigating it in large craft. Captain Bonneville sent a party from Green River in 1833 to make its circuit, but they gave it up on striking the desert west of the lake, lost their way, and after a devious pilgrimage found themselves at last in Lower California. Until Colonel Fremont visited the lake in 1842, on his way to Oregon, it is probable that its heavy briny waters had never been disturbed or the solemn quiet of its mountain-islands broken by man. He pulled out from near the mouth of the Weber River in a rubber boat eighteen feet long for the nearest island, which, when he had climbed and found a mere rock, as he says, fourteen miles in circuit, he named it "Disappointment Island." Captain Stansbury afterward rechristened it "Fremont Island," and such it is called. He found neither timber nor water on it, but luxuriant grasses, wild onions, parsnips and sego. Near the summit the sagebrush were eight feet high and six or eight inches in diameter.

In the early spring of 1850 Captain Stansbury spent three months in making a detailed survey of the shores of the lake and its islands. He found the western shore a salt encrusted desert, in traversing which his men more than once well nigh perished for want of water; the northern shore composed of wide salt marshes overflown under steady southern winds; the Promontory Range, projecting thirty miles into the lake from the north, having many sweet water springs around its base, and a good range (now covered with flocks and herds); the southern shore set with mountain ranges, standing endwise to the lake, with grassy valleys intervening—Spring, Tooele and Jordan; the eastern shore fair irrigable land. The latter was then already sprinkled with infant settlements, and was producing fifty bushels of wheat to the acre. Almost everywhere land and water were divided by mud flats, across which they were forever dragging their boats and packing their baggage.

The principal islands are Antelope and Stansbury, rocky ridges, ranging north and south, rising abruptly from the lake to an altitude of 3,000 feet. Antelope is the nearest to Salt Lake City, and is sixteen miles long. Stansbury is twenty miles to the westward of Antelope, and twelve miles long. Both at that time were accessible from the southern shore by wagon. Both have springs of sweet water, and good grass for stock. The view from the summit of Antelope is described as "grand and magnificent, embracing the whole lake, the islands and the encircling mountains, covered with snow—a superb picture set in a frame-work of silver." Mention is made of the scenery on the eastern side of Stansbury. "Peak towers above peak, and cliff beyond cliff, in lofty magnificence, while, crowning the summit, the 'dome' frowns in gloomy solitude upon the varied scene of bright waters, scattered verdure, and boundless plains (western shore) of arid desolation below." Descending one day from the "dome," the gorge, at first almost shut up between perpendicular cliffs of white sandstone, opened out into a superb wide and gently sloping valley, sheltered on each side to the very water's edge by beetling cliffs, effectually protected from all winds, except on the east, and covered with a most luxuriant growth of bunch-grass. Near the shore were abundant springs of pure soft water, probably covered by the lake now. There was no sweet water on the western side of the island. Of miner islands there are Fremont, Carrington, Gunnison, Dolphin, Mud, Egg, Hat, and several islets without names. With the ranges enclosing the valley they present water-marks at different heights, one principal one, 1,000 feet above the present lake level, indicating a comparatively recent receding of the waters, either from change of climate or of the relative level of the mountains and the basin.

Nothing (in the popular sense) lives in the water, although a species of shrimp, quite nutritive and abundant, have lately been discovered; but aquatic birds cover the shores and islands in the breeding season, carrying their food from the fresh water streams that feed the lake, or feeding on the larvæ of *diptera*, which accumulate in windrows on or near the water's edge. Captain Stansbury navigated and examined the lake thoroughly, and was often, he says, oppressed by its solitude. The lake covers an area of about 2,100 square miles, and has an absolute altitude of about 4,250 feet. Its mean depth is about fifteen feet; the deepest place—between Antelope and Stansbury—is said to be sixty feet. The water is of a beautiful aqua-marine hue, and so clear, when still, that the bottom can be seen through four fathoms. The two principal islands were accessible from the southern shore when the valley was first settled, and the lake then had an area of but 1,700 square miles. After the settlement the water gradually grew deeper, until the storm line was eight or ten feet higher than before settlement. Of course it encroached upon the flat shores and increased its area more than fifty per cent. In recent years it has gradually become shallower again, until now it is but four or five feet deeper than in 1847. Beach marks on the mountain show that the lake has been a thousand feet deeper than it is now, and it was then about ten times as large as it is now. Twice, Mr. Gilbert says, has it been at this high stage, and possibly it may be again, but hardly in our time. *That* lake, as big as Lake Huron, is called Lake Bonneville. Receiving fresh water constantly and having no outlet, the water has become dense and salt. The ocean has about 3.5 per cent of solid matter; Great Salt Lake has 14 to 17 per cent; the Dead Sea, of Asia, 24 per cent. The solid matter varies somewhat in proportion with the seasons, and with dry and wet years, but the range is probably fairly stated above, 14—17. Captain Gilbert estimates the evaporation from the lake every summer at eighty inches. Since about seventeen inches falls upon its surface, the streams must carry in upwards of five feet.

Within the past fifteen years the lake has steadily grown in interest as a watering place. In the long sunny days of summer the water becomes very warm. It is so dense that it sustains a human body without effort. All one has to do is to keep it outside of him; and to swim, one has only to make the proper motions. A more exhilarating exercise than buffetting the waves, when roughened by the wind, it would be hard to imagine, for one must keep it entirely out of him if he would not be seriously distressed, perhaps disabled. This bathing is irresistibly attractive. Between the stimulating effect on the skin, the saline air, and the play of the muscles of every part of the body involved in swimming, one exercises and rests at the same time. It seems to be invigorating, tonic, healing, health-giving, strength-renewing. The accommodations are excellent, so far as they go; 300 additional bathing houses are now going up at Garfield, and much expense has been incurred to make the bathing resorts pleasant, Dancing pavilions have been built out on piles over the water, and here, during the season — May 15 to September 15 — bands are always playing. If people don't want to wet themselves, or to watch the amphibians disporting in the water (which is a sort of " temple of truth "), they can waltz. There are restaurants, bars, play-grounds, booths, games. Boating clubs have been organized, and at a regatta in 1888 a mile and a half with a turn was made in eight minutes and thirty-six seconds, beating the record by thirty-four seconds. The bathing resorts are favorite places for celebrations, banquets, and holiday observance. But we will let Mr. Jones's "Salt Lake City," picture Garfield Beach at length.

GARFIELD BEACH.—The nearest point to the lake from Salt Lake City is about ten miles distant, but it is almost inaccessible on account of the boggy character of the ground. Twenty miles from the city is the great resort known as Garfield Beach, reached by the Union Pacific Railway, and here the shore is sandy and wholesome, abounding in fine retreating bays that seem to have been made on purpose for bathing.

Here the northern peaks of the Oquirrh Range plant their feet in the clear blue brine, with fine curving insteps, leaving no space for muddy levels. The crystal brightness of the water, the wild flowers and lovely mountain scenery, make this a favorite summer resort for pleasure and health seekers. Numerous excursion trains run from the city, and parties, some of them numbering upward of a thousand, go to bathe, and dance, and roam the flowery hillsides together. The railroad carried 100,000 persons to this resort last season. The hotel and bath houses which form the principal improvements of the place cost upwards of $100,000, and are located on a beautiful sandy beach. The station is a building 350x50 feet and twenty-five feet high, furnished with an excellent dining hall, lunch counter, bathing suit office, and an open waiting room, situated about thirty-five feet above the water, giving an extensive view of the lake, and a full view of the pavilion and bathing below, while affording perfect enjoyment of the cool breezes from the water, and protection from the sun. The tower in the center of the building has a second story with an observatory, where a still better view can be obtained. Across the track is the bowery, a commodious building, where people can eat their lunch and enjoy themselves generally. Near by are the games and shooting gallery, and farther off the race track and ball grounds. Two broad stairways lead down from the station to the promenade, on either side of which are 300 bath-houses, each six by eight feet, furnished with wash-stands, shower-baths, mirrors, etc. From the bath-house platform, stairs lead down to the water resting on a beautiful clean sandy bottom, and gradually deepening till beyond the pavilion the depth is sufficient to suit the most exacting. The promenade is about fifteen feet wide and 300 feet long, leading from the stairs at the foot of the station out over the water; it has three towers, is open all around, has a waxed floor for dancing, has innumerable chairs for the accommodation of the public who wish to sit and enjoy the cool breezes and watch the bathers and dancers. Beneath the pavilion and connected with it by a stairway is the steamboat landing, where for twenty-five cents a ride on the lake can be obtained. Near the landing is the boat-floor, where all sorts of boats can be hired. When the lake is calm the bathers can stretch out on their backs and lie as motionless as logs upon the water, and even go to sleep floating around lazily, sometimes spending hours there without danger; but when a storm is on the lake, then the breakers roll in at Garfield, foaming, boiling and pounding on the shore; then the strongest bather can have all the battling with the breakers that he wants, and more than he wants, for the waves come in from the deepest part of the lake with a sweep of seventy-five miles before they break upon the shore.

Within what appears to be a stone's throw from the station rise the lofty Oquirrh Mountains to a height of nearly a mile above the lake; the tall trees near their summits look like brush, and the patches of snow still lingering in the hollows are in strange contrast with the crowds of people cooling off in the warm waters of the lake below. Suppose we take a climb up the mountain to see what we can see. Leaving the bathers to enjoy themselves we start out for the mountains, thinking to reach the base in a minute or two and to reach the top in an hour. We are ten minutes in getting to the base of the mountain, and as we turn to look at the station we find it has grown quite small, and we are at least 100 feet above it; then we climb a few minutes and get out of breath, when sitting down we enjoy the increasingly beautiful views; perhaps in half an hour we get up to what from below seems a roadway along the mountain, but it turns out to be one of the old beaches of the lake; it is almost level, from thirty to 100 feet wide, and as smooth and even as though it were a railroad grade just finished; we could walk along this for miles and find places where it is 300 feet wide; near by we find a large cave which was beaten out of the rocks by the waves of the old lake. It is thirty or forty feet deep and ten to fifteen feet high; here we stop for rest, for the climb is very tiresome. Sitting on a projecting point of the rocks we look down upon Garfield, now reduced to a mere toy, the music of the band playing for the dancers floats up to us in far away tones; the throngs of people swarming around the buildings are reduced to diminutive proportions, and the heavily loaded train just coming in puffing and whistling adds still another thousand to the crowds already there. Before us lies the lake in all its beauty, with its many islands plainly visible even to the farther shore, and all the mountain ranges for many miles on all sides stand out in bold relief. About us are strange and beautiful flowers in great variety. The setting sun reminds us that our time is spent, and as we look up the mountain we seem scarcely to have begun to climb it, so we return to the station and are soon speeding along in the twilight to the city.

Garfield Beach received its name some dozen years ago, from a second notable visit to Utah of our martyred President, who was, it is said, first nominated to the first presidential office by a party of gentlemen and ladies with whom he was making a trip on the lake on board the "City of Corinne," and which was changed to "General Garfield" in further commemoration of his visit. During the hot months cheap trains leave for the bathing grounds daily at the close of business. The run is made in thirty minutes, and the excursion, aside from the bathing, is pleasant. Some day this shore will be built up with private watering place cottages, plentifully interspersed with large, airy hotels, with water and trees for grounds, and it will be thronged in the bathing season as no ordinary seaside resort ever is, for it offers unparalleled attractions in the way of rest, comfort, saline air, and the most delightful and invigorating exercise, calling into play all the muscles; never tiring, the water is so buoyant; never chilling, it is so warm; recreating and invigorating; an incomparable tonic.

EXCURSIONS.—From Salt Lake City there are numerous trips taken by tourists over the Union Pacific System, into Idaho, Oregon, Washington, and to the wonderful Yellowstone Park. Nowhere on the globe is there to be found such a variety of climate, scenery, and resources as between the Missouri River, or the ninety-sixth meridian, and the Pacific Ocean; and in the magnificent stretch of country are found resorts which can be enjoyed at all seasons of the year. The best climate of every known country can be found in this area. Here nature not only equals but excels everything that she has done for mankind in other portions of the globe; and American enterprise and skill have made them accessible to the nations of the earth. From Portland, magnificent ocean steamers depart for the far distant Orient. Fine steamers also ply over the broad bosom of the Pacific Ocean from Portland to Alaska, that wonderful Territory of the north. The Oregon Railway & Navigation Company's steamers, which compare favorable with the Atlantic steamships, make regular trips twice a week from Portland to San Francisco. During every excursion season many thousand tourists visit Alaska.

LAKE PARK.—This resort is within twenty miles of Salt Lake City, near the line of the Union Pacific Railroad to Ogden. The grounds here are nearly level and but slightly above the water. The main building is the pavilion, sixty feet square, and nearly as high; it is an open affair, and, like that at Garfield, has a waxed dancing floor, with seats around the sides, and a platform for the orchestra. There are two buildings, one on either side of the pavilion, 50x30 feet; the one on the north is used as a restaurant, where very good meals are served for 50 cents. On the east side of the track and inside of the loop is the Bowery, fitted up with seats and furnished with a lunch counter, ice water, etc.; near by are the ball, croquet, and tennis grounds; just outside the loop on a grassy slope is a row of summer cottages for visitors; not far away are little arbors tastily fixed up; there is also a large cook-stove, where those so inclined can make their own coffee or tea or cook what they may desire. On the west side of the pavilion are rows of bathing-houses facing westward and parallel with the shore. The water is shallow and the waves are seldom boisterous; there is scarcely any difference in the bouyancy of the water between here and elsewhere. A long covered pier runs out from the shore into the water and is furnished with seats the entire length; it is delightful to spend hours on this pier to enjoy the lake breezes, and watch the bathers and boats that all day long are coming and going, intent upon their own enjoyment.   Since the water here is generally so smooth, rowing is one of the favorite pastimes; and yet this place is almost always favored with a gentle breeze, sufficient for sailing, but not strong enough to make the water very rough; this is due doubtless to the shallowness.

SYRACUSE.—A branch of the Union Pacific runs from Syracuse Junction, nine miles south of Ogden, six miles down to the shore of the lake, and this is another bathing resort, and at the same time the scene of the most extensive and perhaps best equipped salt-making plant on Great Salt Lake.   The entire process may there be seen to best advantage.   The place is made pleasant by groves of trees of large size, put out thirty or forty years ago.   Like the entire east shore, this place is too flat and the bottom too muddy for fine bathing.   It is now proposed to construct a gigantic swimming bath and pump the water of the lake into it for bathing purposes.   This, if it proves satisfactory as a bathing place, will probably be done at Lake Park, which suffers from the same drawbacks.

THE WASATCH CAÑONS.—The Wasatch Mountains and the High Plateaus, like other great chains, are composed of parallel ridges inclosing lateral streams.   Some of these run long distances within the chain before finding a way out.   All of them escape, finally, by making their way through the obstacle, and this breaking through makes the cañon.

WEBER CAÑON.—The divide between the waters of the Rio Colorado and of the Great Basin is crossed by the Union Pacific Railway at Reed's Summit, 7,463 feet above the sea.   Descending a few miles it crosses Bear River at an altitude of 6,969 feet, here flowing generally northward, following it down ten miles, leaving it 6,656 feet above the sea, thence surmounting Echo Pass, 6,785 feet in height, it begins the direct descent into the Great Basin through Echo and Weber Cañons, crossing Weber River at an elevation of 5,240 feet, and striking the level of Salt Lake at Ogden, 4,290 feet.   Echo Cañon is no cañon in the true sense.   A wall of sandstone rises perpendicularly on the north 300 or 400 feet; on the south there is no wall and little rock, but a succession of grassy ridges, sloping smoothly toward the stream.   The road strikes Weber River, another northward-flowing stream, about midway of its course, and follows it down through a valley for five or six miles below Echo City, to the "Thousand Mile Tree," where the mountains draw together, and the first cañon commences.   The valley suddenly narrows to a gorge, the rended rocks tower to the sky and almost overhang the train   Through tunnels and over bridges this cañon is cleared in half a dozen miles, the mountains recede again, and soften down into mere hills in comparison.   An oval valley like the one above is passed, the mountains again close in on the river, and the train enters Devil's Gate Cañon, where the naked rocks rise a mile in the air.   Ages ago they presented a fixed rock dam, which it seems the river could never have conquered, but it has, and, through the passage made by its persistence, the road soon emerges from Devil's Gate into the summery airs of the valley.   The scenery has been described and illustrated until the traveling public is familiar with it. But one gets only a slight idea of its beauty and grandeur from a ride through it on the railroad. He must stop off, and, on foot or horseback, explore the side streams and reach various elevations half a mile above the river, before he can be said to have seen it all.

Bear River flows along a distance northward before it finds a passage outward into Cache Valley, and thence into Salt Lake Valley. Similarly the Provo, rising near the source of the Weber, and flowing southward, has its Alpine valleys, and finally cañons out into the Basin. So with the Sevier, and its affluents, and so to a less extent the various minor streams that flow westward into the Basin directly from their sources, as Logan and Blacksmith Forks, Box-Elder Creek, Ogden River, the Cottonwoods, American and Spanish Forks and many others. One can see something of them in hastily passing through them, but to get the full benefit he must have a camp outfit, his own conveyance and time, saddle-horses, hunting and fishing tackle, and all the paraphernalia of the sightseer, the tourist and the sportsman. For such it is hard to select the localty, since the Wasatch Range affords such an endless variety from end to end.

CACHE VALLEY.—Cache Valley, which is traversed from end to end by the Union Pacific Railway, is an inviting field for the tourist. It is literally cached among the ridges of the Wasatch, like San Pete, Ogden, Alpine, Morgan, Echo, Rhodes and Sevier Valleys; and is as though round a symmetrical oval area, ten by fifty miles, the mountains had risen or ranged themselves at some mysterious bidding, to show what could be done in the way of valley making. It is about 4,500 feet above the sea, well watered, inclosed by mountains 8,000 feet high, in whose gorges the snow lies until August, their shaggy sides meanwhile invaded by the green of the valley, which creeps to their summits between the snow banks, or appears in sunny places among the scattered pines and dark points and ridges of rock. A fair sprinkling of forest would perfect the picture, but this it lacks, and the green of the valley and mountains only relieve the eternal gray-brown of everything else after all. The range on the east is the main Wasatch, deeply notched by the streams, which are alive with trout, and afford passage over fine roads to Bear Lake Valley, fifty miles eastward. Where the rivers emerge from their cañons and rush laughing into the sunshine, there waters are caught up and led in a thousand trickling rills to bless the fields with fatness. Some lighter streams and springs perform the same kindly office for the west side, and so there is a belt of cultivated land sprinkled with towns all around the edge of the valley. Of these Logan is the largest, and Smithfield the prettiest. From several points on the railway Cache Valley is a lovely sight. One can drive on fine roads all around it, to Soda Springs and Bear Lake, and over hill and dale southward into Salt Lake Valley via Box-Elder Creek or Ogden River.

OGDEN CAÑON.—The same section may be penetrated almost as well from Ogden. Of the interesting places in the immediate vicinity of Ogden, the cañon of Ogden River ranks highest. There is a good carriage road through the cañon, which is ten or twelve miles long, and the pass-age presents the same immense close towering rocky walls, broken apart by the full roaring stream, common to all the cañons of the Rocky Mountains. Power of resistance on the one hand, and of attack on the other, are well symbolized. There are minerals and mineral springs along the way. Through the outlaying range one enters Ogden Valley, an enclosed park, with its settlements and farms, beyond which the drive extends into both Bear Lake and Cache Valleys. All the streams in that part of the Territory afford good sport for the angler, and the valleys and hills are grass grown and alive with grouse and snipe, sage hens and prairie chickens.

PARLEY'S PARK.—From Salt Lake City, Parley's Park, the Big Cotton-wood Lakes, and American Fork Cañon are the favorite resorts. The park is about twenty-five miles from Salt Lake City, just over the crest of the Wasatch, on the sources of the Weber, and nearly as high as the mountains themselves. The road ascends through Parley's Cañon, and is a fine drive. There is a hotel in the park, but visitors usually prefer taking along with their team their own camping outfit. The elevation insures refreshing coolness, especially of the nights. The park is quite extensive in area, affording good drives, fishing and hunting, and stretches for horseback riding, and among other objects of interest, Park City and the Ontario mill and mine. One can get a fair idea of the ways and means of mining by a visit to this town, mine, and mining district. Excursions may be made eastward to the sources of the Weber and Provo Rivers, the whole region being full of interest. The country-side, an old formation, apparently, giving evidence of the mighty action of water or ice, or both, geological ages ago.

BIG COTTONWOOD LAKES.—There are a series of small lakes at the head of Big Cottonwood, at the most picturesque of which, namely Mary's Mr. Brighton has built a hotel for the accommodation of summer visitors. For many years it has been a mountain resort, and the number of persons seeking its cool fresh air, and the enjoyment to be derived from a study of nature, is yearly increasing. The hotel is always full during the hot months, and the lake bordered all around with the tents and wagons of campers. Excursions must be made on foot or horseback. They may include visits to Park City, Heber City, Midway, or Kamas, to the Big and Little Cottonwood mines, to other rock-bound tarns, and to sightly peaks. From these one can look out over Jordan Valley, the lower section of the Oquirrh, Rush Valley, and in clear weather upon the far summits of the Deep Creek Mountains, glittering like silver points in the distance. Perhaps the finest view is from Bald Peak, among the highest of the range. Standing on its top, 20,000 square miles of mountain, gorge, lake, and valley, may be swept by the eye. Sixty miles south, Mount Nebo bounds the view. Beneath lies Utah Lake, a clear mirror bordered by tawny slopes, and Salt Lake City, embowered in foliage, with Great Salt Lake rolling its white caps and glittering in the sunshine beyond, its islands and all the valley ranges dwarfed to hills. Northward the higher points of the Wasatch catch the eye until they are lost in the distance. Eastward the sources of the Weber and the Provo fill the foreground, while successive mountain ranges bound the view in that direction. Words can give but a faint idea of the magnificence of the outlook from Bald Peak, or Kesler's Peak, or Mount Clayton, the latter the corner of three counties, and from whose bare sides start Snake Creek, the Cottonwoods, and American Fork, or any other of the higher summits in the vicinity of Mary's Lake.

AMERICAN FORK CAÑON.—South of the Cottonwoods, American Fork Cañon opens into the Utah Lake Basin. It has been called the Yosemite of Utah, and undoubtedly its succession of wild gorges and timbered vales makes it the most picturesque and interesting of any of the cañons of the Wasatch. To visit it now one must take horse or carriage at American Fork, thirty-two miles south of Salt Lake, and the better way will be to take along a complete outfit for camping, although there are buildings at Deer Creek and at Forest City. At Deer Creek one takes horses, eight miles to Forest City, and then the ascent to the Miller mine, or the Silver Bell, begin. It is four miles farther, the mines being 11,000 feet in altitude. Once there, it is but a short climb to the top of the peak, nearly as high as any of the range, and affording a most magnificent and almost unbounded view in fine weather.

This cañon is noted, not only for the towering altitude of its enclosing walls, but for the picturesqueness of the infinite shapes, resembling artificial objects, towers, pinnacles and minarets chiefly, into which the elements have worn them. At first the formation is granite and the cliffs rise to a lofty height almost vertically. Then comes quartzite, or rocks of looser texture, conglomerates and sandstones. The cañon opens only to the sky, and you enter a noble gallery, the sides of which recede at an angle of 45° to a dizzy height, profusely set with these elemental sculptures in endless variety of size and pattern, often richly colored. "Towers, battlements, shattered castles, and the images of mighty sentinels," says one, " exhibit their outlines against the sky. Rocks twisted, gnarled and distorted; here a mass like the skeleton of some colossal tree which lightning had wrenched and burnt to fixed cinder; there another, vast and overhanging, apparently crumbling and threatening to fall to ruin." At Deer Creek the cañon proper ceases; the road has climbed out of it, 2,500 feet in eight miles. This is the main resort of pleasure parties. The wagon road continues to Forest City, eight miles above. The surroundings are still mountainous, but there are breaks where the brooks come in off grassy hills and forests of aspen and pine. Forest City has been a charcoaling station for years.

To the sublimity of cañon scenery in summer, an indescribable beauty is added in the autumn, when the deciduous trees and shrubbery on a thousand slopes, touched by the frost, present the colors of a rich painting, and meet the eye wherever it rests. To get the full benefit of this one must go up and up till there is nothing higher to climb. In winter another phase succeeds. The snow, descending for days and days, buries the forests and fills the cañon. Accumulating on high and steep acclivities, it starts without warning and buries in ruins whatever may be in its track. Hardly a year passes that miners and teamsters, wagons and cabins, are not swept away and buried out of sight for months. The avalanche of the Wasatch is as formidable as that of the Alps. Probably forty feet of snow falls on the main range every winter. Seven miles of tramway in Little Cottonwood Cañon are closely and strongly shedded for defense against snow-slides. Even this is not always effectual. Yet the main traveled roads over this range, whether wagon or railroads, are but slightly obstructed by snow as a general thing.

UTAH BASIN.—Utah Basin is shut off from Salt Lake Basin by a low range cut through by the Jordan River and run through by the Union Pacific and Rio Grande Western Railroads. Its prettiest feature is a sheet of sweet water, twenty-five miles in length and about five in breadth, with broad grassy slopes from the water's edge to the foot of enclosing mountains. It receives the American, Provo, and Spanish Rivers, and discharges into Great Salt Lake through the Jordan River. It abounds in fish, principally speckled trout, of large size and good flavor. This made it a resort of the Utah Indians in former days, after whom the lake, the county, and the Territory seem to have been named. It is a pity the other Indian names of springs and creeks in this lovely basin have not been likewise preserved—Timpanogas, Pomontquint, Waketeke, Pinquan, Pequinnetta, Petenete, Pungun, Watage, Onapah, Timpa, Mouna, and so on. They have all been superseded, and their memory is fast passing away, as the Indians themselves have done.

"On the Timpanogas (Provo) Bottoms," said Lieutenant Gunnison forty years ago, "wheat grows most luxuriantly and the root crops are seldom excelled. A continuous field can be made thence to the Waketeke (Summit) Creek and the lovely Utah Valley made to sustain a population of more than 100,000 inhabitants." The field was long since made, and the population now numbers nearly 30,000. The leading town is Provo on the Timpanogas, under the overshadowing Wasatch. It is like all the better class of towns in Utah, regularly laid out, an accumulation of garden spots, the houses half hidden by the foliage of fruit trees and vines. Provo is about fifty miles south of Salt Lake City, and is a good outfitting point for the tourist. The principal attractions in the vicinity are Utah Lake and Provo river. The latter has the inevitable cañon, above which a good wagon road leads through a succession of settled Alpine valleys to Kamas Prairie, which Captain Stansbury describes as "a most lovely, fertile, level prairie, ten or twelve miles long and six or seven miles wide," where the affluents of the Provo and Weber interlock. The drive may proceed down the Weber to Ogden, if one desire, with the same alternation of landlocked valleys and mountain gorges. A dozen thriving settlements will have been passed through en route.

Six miles south of Provo is Springfield, where, by road up Spanish Fork Cañon, we proceed into the finest timbered, tallest grassed, best watered section of Utah, presenting a fresh field for hunting and fishing. All along here the Wasatch Range presents a most interesting aspect, and frequently offers access via cañons of more or less attractiveness. An isolated range trending north and south, west of Utah Lake, divides the basin into separate halves, cutting off Cedar and Goshen Valleys (dry for the most part, and of little account) sloping gradually up for twenty miles to the summit of the Oquirrh, 6,000 feet high, on the western side of which are the Tintic mines.

On the Sevier.—Utah Lake Basin may be said to end in the vicinity of Nephi, under Mount Nebo, where Onapah (Salt Creek) Cañon opens the way for another side railroad into San Pete Valley, with its eight or ten settlements and 10,000 to 12,000 inhabitants. From the head of San Pete one may find his way northward into Spanish Fork, or eastward over a mountain into Thistle or Castle Valleys. Southward the valley opens on the Sevier River, a world in itself, with passes of majestic grandeur through ranges on either hand into adjoining valleys. A journey up the Sevier in fine weather is very interesting, and so is the region about its heads, where the waters divide and flow apart. In his "Geology of the High Plateau," Captain Dutton often becomes eloquent in the eff rt to word-paint the scenery of that region. Panquitch Lake, at an absolute altitude of at least 8,000 feet, a mile and a half wide by a mile and a half long, is described by him as "a delightful locality both for the tourist and the geologist." The shores and slopes are wooded and the floors of broad and stately ravines bearing sparkling streams are carpeted with long rich grasses, and every knoll and sloping bank is a parterre of the gayest flowers. Fish Lake is a much larger body of water—5½ by 1½ miles—8,600 feet above the sea, walled in by two noble palisades, respectively 1,600 and 2,600 feet high. "No resort more beautiful than this lake can be found in Southern Utah." The outlet is into the Colorado, but it was formerly into the Sevier, and might be recovered by a short tunnel. The Union Pacific runs a branch from Nephi through Salt Creek cañon into San Pete Valley, and the Rio Grande Western, starting from Thistle Station in Spanish Fork Cañon, are operating a line to Manti, and are building on south to and up the Sevier. The building of railroads promises to open the upper Sevier region to the tourist and sportsman as well as to business and commerce.

In the Uintahs.—The following is from Mr. M. E. Jones's "Salt Lake City:"

"To the artist and hunter wishing to see nature in all her native wildness, there are few places superior to the grand old Uintas, the loftiest mountains of Utah. These are reached along the Weber or the Bear, both on the Union Pacific. Going to Wanship by rail, we there hire a a team to take us to the mountains; we can get to our destination in a day, but for pleasure's sake we take short stages, and at almost every camp spend hours in fishing, catching all we can carry almost every time we go out, and the finest trout, too. The valley of the Weber is a broad one nearly all the way to the Uintah. We ascend gradually until we are among the sub-alpine meadows nearly 9,000 feet above the sea. Here are aspens in abundance and groves of majestic spruces, beautiful grassy plats under the trees, beautiful lakes filled with fish, sparkling cascades and waterfalls, rocks and cliffs, fallen timber, the finest fuel, and all sorts of game both large and small. The flowers are fully as plentiful as in the Wasatch, and there are many new kinds. The grouse are very abundant; one need not go out a day without bringing home a deer; there are elk and bear. Here it freezes every night in the year, but the days are warm and pleasant; the skies are clear, but with an occasional summer shower. The great peaks are some ten miles off, rising gray and bleak against the sky.

"The most enjoyment will be found in going into the Uintahs at the head of the Bear, for there we get amongst the great peaks at once. The Bear is reached from Hilliard or Coalville, at first a broad open valley, then a succession of knolls and hollows, and groves of spruce and fir and pine carpeted with flowers. At last the base of the great wall of peaks is achieved, and alpine meadows, carpeted with luxuriant grass and decked with wild flowers of every hue. On both sides of us the great U-shaped valley rises several thousand feet, clad with dark forests. In front, the immense peaks with their attendant walls tower into the sky nearly 14,000 feet above the sea. all lighted up with glistening snow. We are at an elevation of 9,000 feet above the sea, where we can lounge around, fish, hunt, sketch, and study nature in all its phases. The geology of this place is intensely interesting, since here the glaciers held out the longest, and the evidences of their existence are as fresh as though they had melted but yesterday. The fishing is fine, and deer, grouse, and elk are plentiful, and bear can be found by seeking them. The wealth of flowers is fully as great as in the Wasatch, while there are many kinds not found in any other mountains of Utah. The scenery here is vast, grand, and, because of the work of the glaciers, destitute of narrow gorges and rugged cañons except at the heads of the streams, where all the magnificence of the Wasatch is multiplied till the views produce the sensation of sublimity and bewilderment. If we stand upon the summit of La Motte Peak, over 13,000 feet above the sea, we are upon a narrow ridge above the clouds; a single step would precipitate us thousands of feet before we reach the bottom; the lofty trees so far below appear like tufts of grass; clouds float lazily beneath us; and through the rifts we see the silvery threads where the cataracts are flowing, but no sound comes to our far height. Around us on all sides rise massive cliffs and precipices thousands of feet high, vast beyond all comprehension, and yet so well proportioned that they remind us of spires, castles, domes, cathedrals, and amphitheatres, cut out of the rock by a giant bygone race. In the midst of the amphitheatres lie the beautiful, shining strings of pearls, the alpine lakes, the last resting places of the mighty glaciers that perished there. Long ago these plowmen, with overwhelming force, cut up the narrow cañons into broad and fertile valleys, now covered with luxuriant grass and groves of trees, the homes of elk and deer. As we look down from our perch among the clouds, we see long lines, a thousand feet high, of rounded boulders which the glaciers left fringing the valleys on either side. Near the heads of the valleys we behold a series of massive embankments crossing them at right angles, forming beautiful lakes, as if the dying glaciers attempted to stop the rushing waters, and at each failure formed new dams higher up; and so on, till the last embankment lies at the very head, like a wall of freshly broken stones piled with great regularity and care, and still but half done, as if the glaciers had died but yesterday toiling at their tasks. As we look off over the peaks, we see an immense stretch of country. On the north the valley of the Bear lies spread out at our feet, we can see the Union Pacific Railway twenty miles away, and range upon range of mountains for at least 100 miles beyond. On the west, Reed's Peak is in the foreground, towering hundreds of feet above us, with its masses of unmelted snow, its black beard of fringing forests, its green lakes, and silvery threads of water flowing from them. Farther off, nearly seventy-five miles away, rise the Wasatch peaks, and we can even discern the Oquirrh Mountains beyond. On the southwest, beyond the Uintah peaks, are many mountain ridges as closely compacted together as the backs of animals in a herd, and far away on the horizon 150 miles off rises the camel's hump of Mt. Nebo, gray and hazy, but still plainly visible. On the south we look over into many parks and can almost see the deer and elk feeding there in places almost untrodden by the foot of man; there is the head of the Duchesne; not far off head the Provo and Weber; and at our feet the Bear starts on its northward way. East of us continues the great Uintah Range with peaks, a number of them higher than our own, and all rising far above the timber line, cold and bleak, with great masses of glistening snow, and yet at this time gorgeous with alpine and rare flowers, except on the very summit where are only piles of huge stones. How far we can see it is difficult to tell, but our horizon is not less than 200 miles in diameter. No one can ever appreciate the vastness of this country until he ascends one of our lofty peaks, and by the assistance of our remarkably pure air sees as far as the rotundity of the earth will permit. and that too at an elevation of nearly two miles above the face of the country."

CONCLUSION.—During the past two or three years Utah has in some way been stirred with new life; the more central parts have advanced at a marvelous pace; realty in all the towns of Salt Lake and Utah Valleys has doubled and doubled again in price; the output of the mines has largely increased; narrow-gauge roads have been standard-gauged for hundreds of miles; new roads have been projected and have struggled to their feet; lines have been consolidated and are being extended in fruitful directions; building in the towns has exceeded all former expectations; and public improvements, long neglected, are started easily and pushed with vigor.

Statehood, so long promised, is now near at hand; and that Utah, with her great, growing, intelligent and thrifty population, her magnificent resources developed and to be developed, her glorious climate, her fertile soil, her health resorts and her natural attractions, will be one of the brightest stars in the galaxy, can no longer be disputed.

www.ingramcontent.com/pod-product-compliance
Lightning Source LLC
Chambersburg PA
CBHW021821190326
41518CB00007B/692